D0049557

THE
DARWIN
MYTH

C. 1

THE
DARWIN
MYTH

THE LIFE AND LIES OF CHARLES DARWIN

BENJAMIN WIKER, PH.D.

Author of *10 Books That Screwed Up the World*

Since 1947
REGNERY
PUBLISHING, INC.
An Eagle Publishing Company • Washington, DC

Cataloging-in-Publication data on file with the Library of Congress

ISBN 978-1-59698-097-6

Published in the United States by
Regnery Publishing, Inc.
One Massachusetts Avenue, NW
Washington, DC 20001

www.regnery.com

Manufactured in the United States of America

10 9 8 7 6 5 4 3 2 1

Books are available in quantity for promotional or premium use. Write to Director of Special Sales, Regnery Publishing, Inc., One Massachusetts Avenue NW, Washington, DC 20001, for information on discounts and terms or call (202) 216-0600.

Distributed to the trade by:
Perseus Distribution
387 Park Avenue South
New York, NY 10016

*To my beloved children, Jacob, Anna, Faith,
Clare, Nathaniel, Beatrice, and Rachel.*

Contents

Introduction

The Darwin Myth

The year 2009 has been dubbed the Year of Darwin, because it is the 200th anniversary of his birth and the 150th anniversary of the publication of his *Origin of Species*. It is high time we understood who Darwin really was, and what he really did. Distinguishing the facts from fancies is no easy matter for two related reasons. First, Darwin himself is often positively misleading about his own life, and not just because he had trouble sorting out names and dates as an old man penning his *Autobiography*. Second, biographers of Darwin have too often taken him at his word when they should have exercised a little more skepticism; moreover, they have tended toward hagiography, making Darwin a kind of secular saint who singlehandedly brought enlightenment to a world shrouded in the darkness of superstition and ignorance. In one sense, they can hardly be blamed. That's how Darwin wanted to see himself, so that's how, in his own humble and ingratiating way, he presented himself to the world. So it was that Darwin and Darwin's biographers have created a myth where there should be a man.

I don't mean to say that Charles Darwin was a bad man. In fact, he was a very good man, and it is part of my task in this book, a pleasurable part, to offer a vivid portrait of one of the most likeable, congenial, self-effacing, patient men of science; a model husband and father, kind and loving, generous and humorous, magnanimous and solicitous toward his neighbors of every social rank.

I stress these qualities because there is another tendency in treating Charles Darwin, a demonizing rather than canonizing urge. This tendency arises in part from the ill-effects of Darwin*ism*, real or assumed. The influence of Darwinism upon Western civilization is immeasurably great. We entered the nineteenth century with Christian assumptions for the most part intact: that we were fallen but redeemable creatures made in the image of God. We exited in a godless cosmos, as mere animals who had managed, through much luck and struggle, to climb from unimaginably low origins to a little above the apes. That news was shocking enough, and it resulted in a kind of reactionary zeal to attack the messenger as the very devil. But no one who met Darwin himself, who really got to know him, could think him a demon. He had too many of the natural, personal qualities of a saint, and in fact, had he not been so entirely bent on creating a godless account of evolution, he might, just might, have become one. God only knows. But certainly the theory of evolution would have been a lot better for it.

I know that seems a rather odd thing to say. The problem with Charles Darwin is not evolution itself, but his strange insistence on creating an entirely *godless* account of evolution. That evolution *must* be godless to be scientific is the Darwin Myth, so profoundly misleading that it must be called a great lie, one that is unfortunately at the heart of his life and legacy. I cannot ultimately explain why Darwin himself so strongly, so implacably insisted on evolution being entirely incompatible with belief in God—although I will offer several important clues and contributing factors. But no mere biographer can read the innermost depths of a soul, least of all, that of a man long dead.

Darwin's insistence that evolution be godless is the cause of much mischief and not a little mayhem. As we will see, some of Darwin's most trenchant critics of this strange insistence were also his best friends and allies. It is a myth that evolutionary theory must coincide with Darwinian theory. It is a myth based on Darwin's fame, but it has distorted our understanding of the scientific evidence and the debates about it.

Darwin's triumph has been to set ideological atheism as the default position of science; as the prism through which scientists are supposed to see the world and conduct their work. It is just as distorting to science as ideological Marxism is to the study of economics. It offers an answer for everything; it is an answer to which facts are twisted to conform; but it might be the wrong answer. Casting Darwin as the apostle of light leading us from a path of superstition has had the unfortunate effect of ruling out of order, as sheer reactionary ignorance, any questioning of whether Darwin might be leading us down another, opposite path of superstition. What is certain is that Charles Darwin, despite his fine personal qualities, was dishonest in this regard, and Darwinism consequently makes for bad science however illuminating it is in regard to many of the details of evolution.

But the problem with Darwinism is not just science. As we will soon see, Darwin's intense desire to set forth a God-free view of evolution brought him to offer an account of human development in which everything about human beings, even their moral capacities, is explained entirely as the result of natural selection, that is, of the struggle for survival where the more fit eliminate the less fit. So-called "social Darwinism" is not, as is typically assumed today, a misapplication of Darwinism, it *is* Darwinism, and it provides an open rationale for eugenics and racism. This had abhorrent consequences in the twentieth century; and unless we understand Darwinism's flaws, there is no reason to believe it will not have equally abhorrent consequences in our own.

Chapter 1

A Very Ordinary Boy

Charles Darwin would change the world with his theory of evolution—only it really wasn't so much his theory as it was his family's theory, going back two generations.

He came from a line doctors. In fact, he was named after an uncle, a physician-in-training at the University of Edinburgh, who had inadvertently cut his own finger dissecting a corpse. The corpse had been "in a state of dangerously advanced putrefaction,"[1] infection set in, and the young Charles Darwin died, not yet twenty.

The more famous Charles Darwin was the son of Robert Darwin, himself a prosperous physician. Charles's grandfather was Erasmus Darwin,[2] who was not only a physician, but a poet, a philosopher, and a propounder of what he called "transmutationism," which was evolution by another name.

Erasmus was the son of a corpulent barrister who had inherited a fine country manor, Elston Hall. He was a towering, celebrated figure, and an eighteenth century man in every respect. Roll the age of Enlightenment into a great ball—Deist skepticism of Christianity, political radicalism, scientific adventurism, a palpitating mercantile spirit, the romance of technology, and a polite Epicurean disrespect for traditional sexual morality— top it with a ponderous head riddled with pox-marked skin and set with penetrating eyes radiating a restless, brilliant, supremely confident intellect; give that head a witty but stammering tongue; and finally place the whole vast frame on legs, one of which was rendered lame by a carriage accident. That was Erasmus Darwin, a man and his age at once; someone who could be compared favorably to his own contemporary and friend Benjamin Franklin.

Like Franklin, he had a restless mind. He had sketched out plans for a steam-driven carriage, with an ingenious steering mechanism, several years before meeting James Watt, the inventor of the steam engine. Over the years he designed a windmill with a third more power than the ordinary model, a machine for lifting boats in canals, and even a mechanical bird. He built a speaking machine (a wooden mouth with leather lips that enunciated "the *p*, *b*, *m*, and the vowel *a*, with so great nicety as to deceive all who hear it unseen, when it pronounced the words *mama*, *papa*, *map*, and *pam*"[3]), and a copying machine that so neatly scribed a duplicate that it was indistinguishable from the original. And that is only a partial list of his technical creativity.

As a physician, he was so well-respected that King George III himself had asked for his services. Erasmus, however, was too much of a Whig—a liberal—to minister to the Tory of Tories. As a man of science, he wrote the *Zoönomia*, a medical-zoological treatise that spelled out his theory of evolution more than half a century before his grandson Charles Darwin wrote *On the Origin of Species*. The *Zoönomia* was a huge international success, with five American editions, three Irish editions, and translations into German, Italian, French, and Portuguese.

Charles Darwin's father, Robert, was the fainter image of the great Erasmus. He had been carefully shaped to take the Darwin place in medicine and Whig society. Charles remembered his father as a large and commanding man, "about 6 feet 2 inches in height, with broad shoulders, and very corpulent, so that he was the largest man whom I ever saw."[4] Robert shared much of Erasmus's wit and his ability to dominate a room (his imposing physique helped). But he rechanneled Erasmus's passion for science and social revolution into the passion for making money and keeping society stable.

Erasmus died before Charles was born, but his influence was great, even if his grandson inherited only his stammer and none of his boisterous charm. Unlike his masterful forbears, Charles did not sparkle. He had no electrifying physical presence. He was just under six feet tall, thick-set when young and lanky when old, at one point later in life weighing less than 150 pounds. When we see pictures of him as an older man, with his characteristic great beard and hoary, beetling eyebrows, he looks much bulkier, but that is the effect of the size of his prominent head, and several layers of clothes and a great coat to keep him warm. As a boy he was a bit chubby, but as a man, he was as thin as his father was fat.

If his grandfather lit up a room with his presence, like a glimmering Christmas tree in the parlor, young Charles was more a comfortable brown sofa set in a darker corner, bulky and nondescript, but loved dearly by those who would sit with him long enough. He never lost this original humility, this feeling of not being the center of attention, of being merely someone who should quietly shuffle in and politely sit down. Upon walking into a great scientific banquet hall when he was old and quite famous, he was rather startled to have everyone look his way and suddenly break into applause. He instinctively turned around to see who had followed him in. It took Darwin some time to realize they were clapping for him.

Charles was not a handsome man; in fact, his contemporaries used rather unflattering adjectives to describe him: bulky, heavy-browed, thick-set, and as he noted of himself, he had a nose as big as a fist. (Captain FitzRoy of the HMS *Beagle*, on which Darwin later sailed around the world, was a casual devotee of phrenology. He thought Darwin's great bump of a nose was a sign of insufficient energy and determination; the captain half-joked that he nearly rejected Darwin as the ship's naturalist because of it.[5]) Darwin's appearance, even as a young man, but certainly as he grew older and sported a great beard, might best be described as simian, which made no end of sport for his detractors later on, especially the cartoonists who, with little ink and effort and much spiteful glee, made him half-ape.

On the positive side, all were agreed that this most controversial of men had no sharp edges to rub against in his personality, but was unfailingly amiable and affectionate, as loyal and loving a boy and then family man as one could ever hope to

find. As a child, Charles was doted on by his older sisters, especially after his mother died when he was only eight. Otherwise,
he had relatively few friends, preferring to stick closely to his
family. However stern his father may have been, his elder
brother and four sisters provided a great nest thickly padded
with affection. Those whom Charles did befriend found him a
hidden treasure. He loved what was familiar, and he was deeply
familiar with what and whom he loved. Even as an old man, the
great and renowned center of controversy, Charles clung to his
wife, his children, and his home, Down House, in Kent, about
sixteen miles from London. It is said that he took pleasure in
drinking from the same old Wedgwood teacup year after year,
the saucer broken and the gilding worn off. Some put Charles's
unwillingness to throw away his chipped teacup to miserliness
inherited from his father. I think it more likely that it brought
him the great comfort of familiarity, an object like an old friend,
worn by daily contact that conforms gently to one's person and
the satisfying rhythms of one's life.

The teacup was a family heirloom, because the Wedgwood
and Darwin families had been allied since his grandfather's days.
Josiah Wedgwood, an extremely successful potter—European
and British royalty were among his customers—was also an amateur scientist and a close friend of Erasmus Darwin. Charles's
father Robert had married a Wedgwood. He inherited a fortune
after Josiah Wedgwood's death, and later, so did Charles. With a
singular exception—his voyage on the HMS *Beagle*—Charles
was not one to stray far from what he knew and loved.

He was very ordinary indeed, and he loved being ordinary.
He loved the ordinary itself. In short, if you saw the boy Charles

Darwin or the young man Charles Darwin, he would have been the last man you would ever pick to be Charles Darwin, the person credited with creating a revolution that shook and is still shaking western society.

Thomas Huxley, who would later become Darwin's bulldog, bully-pulpit preacher, defender, and tireless evolutionary propagandist, would certainly have made a much better Charles Darwin. Huxley was a take-no-prisoners revolutionary, dashing in appearance, electrifying on stage, and armed with a wit that would make a razor dull by comparison. If I were casting history, I'd pick Huxley to play the part of Darwin.

And there was Charles's grandfather Erasmus, the first Darwin who championed evolution. He did so with the unfailing charm of a man simultaneously inebriated by the poetic muse and intoxicated by a (nearly) godless vision of species transforming, one into another, from the first shapeless ancestor, through every variation of every living being, each "possessing the faculty of continuing to improve...and of delivering down those improvements by generation to its posterity, world without end!"[6]

Nothing in the childhood of Charles himself would have appeared to indicate future greatness. He was born on February 12, 1809, the very day that, half-way across the world in a log shack in Kentucky, Nancy Lincoln gave birth to Abraham, a boy with a likewise hidden destiny. Charles was preceded by Marianne, Caroline, Susan, and his best boyhood friend and only brother, Erasmus, and then Emily came along afterward.

Charles emulated and adored Erasmus, who had all of the sparkling qualities Charles lacked. Considered by everyone to be more intelligent and evidently more of a wit, Erasmus seemed a much likelier candidate to carry forward the Darwin name in medicine. Charles was taken to be all too ordinary in intelligence—a mistake as it turns out—and was far more fond of play than school. He loved the outdoors, and in this promised to grow into a good country gentleman, shooting, running dogs, and collecting curiosities from the stream, wood, and field, and generally living a long, happy life on inherited money without any of the burdens of achievement.

He certainly loved to collect things, but then so do many boys. He later mused that the "passion for collecting, which leads a man to be a systematic naturalist, a virtuoso or a miser, was very strong in me, and was clearly innate, as none of my sisters or brother ever had this taste."[7] His grandfather Erasmus had it, for he greedily collected every scrap of scientific and technical knowledge as signs that a new Enlightenment world was dawning, leaving the old world of superstition behind. His father Robert had it, too, but in his case it was for collecting money. But apparently none of his siblings had it. Charles had it for every fascinating tidbit of nature, the more minute the better.

Soon after his mother's death, Charles was sent off to a nasty little boarding school in his hometown of Shrewsbury, in Shropshire County. Boarding school was the usual fate of English boys of his class. Shrewsbury School was almost Dickensian in its melodramatic bleakness. The students were unruly and barbarous, the masters bland and rigorously demanding, and the comfort and hygiene of the boarders were ignored or neglected. Charles once bragged to his sisters, who seemed disgusted rather than impressed, that he washed his feet once a week whether they needed it or not.

It was at Shrewsbury that Charles discovered his inaptitude for math and foreign languages (either ancient or modern). "When I left school," he recalled in his autobiography, "I was for my age neither high nor low in it; and I believe that I was considered by all my masters and by my Father as a very ordinary boy, rather below the common standard in intellect."[8]

Charles was no doubt being unfair to himself. The truth is more likely that his first school experience was dismal. He was away from the comforts of home too soon after his mother's death. For the first time in his life he was immersed in complete discomfort ("20 or 30 boys" stuffed into a dormitory with "only a single window at the end," creating a miasma so vile that the memory of the "atrocious smell of that room in the morning" could still sicken him three score years afterwards[9]). The food was nauseating and the academics stale. But his experiences were far from unique, as many boys marched disconsolately through the deadening halls of similar schools only to come out at the end flaunting their school ties and, in turn, sending their own sons through the very same trial by ashes.

Darwin did find his compensations. One was being introduced to the glories of Euclid—the one aspect of math he understood. He enjoyed reading Shakespeare on his own. He found pleasure in trolling the woods and fields—for nature, he thought, was a far better teacher than the musty old books he was required to read. Above all, he enjoyed opportunities to hunt. "In the latter part of my school life I became passionately fond of shooting, and I do not believe that anyone could have shown more zeal for the most holy cause than I did for shooting birds."[10]

He also enjoyed his own chemistry lab, primitive as it was, that he set up with his brother Erasmus, but this was at home, and therefore something he could use only during his holidays. Here too, the man was in the boy. Charles and Erasmus liked nothing better than amateur chemistry, using scraps of this and that, and whatever bottles and containers could be gathered and nicked from kitchens and shelves to fill out the laboratory as best they could. So passionate was he about chemistry, that his schoolmates nicknamed him "Gas." Many years later, Darwin's own fascinating studies of plants and earthworms, central to filling out the broad contours of his evolutionary theory with the minutest details of nature, were carried on at Down House using bits and pieces of string, kitchen crocks, gardening tools, and every odd thing he could make suit his ends (much to the dismay of the cook and gardener when things disappeared from their domains).

Finally, as with all schoolboys, for Charles there was the thrill of escape, the thrill of *not* being in school, of finally coming home for the holidays, or even of stealing away for a few precious hours. With his home, the Mount, less than a mile away from the boarding school, Charles would dash off in the time allowed before locking up at night, drink greedily of humane, domestic pleasures, and then dash back again, the threat of the bell marking curfew and his expulsion looming ahead of him. On more than one occasion, it brought him as close to sincere prayer as this scion of a distinguished line of freethinkers could get. When in doubt whether he would make it in time, "I prayed earnestly to God to help me, and I well remember that I attributed my success to the prayers and not to my quick running, and marveled how generally I was aided."[11] So Darwin wrote in his late sixties or early seventies, looking back on his bygone days "as if I were a dead man in another world looking back at my own life."[12] He had long since methodically substituted quick running—or swift flying, keen eyesight, strength of limb, or more elaborate or concealing plumage—for every alleged act of God.

His grey imprisonment at Shrewsbury ended in June 1825, when Darwin was sixteen. His father could see no good was coming of it, so he was soon packed off to the University of Edinburgh in Scotland, following his older brother Erasmus, his father Robert, and his grandfather Erasmus into the well-ploughed Darwin family field of medicine.

Grandfather Erasmus had gone to Edinburgh Medical School in 1753. His son Robert had been pressed into the mold even though he had no real interest in medicine. In fact, he hated the sight of blood, even as he loved the sight of money. Nevertheless, because medicine had brought the family wealth and influential contacts, Robert was determined to make doctors of his sons Erasmus and Charles.

When Charles arrived at Edinburgh, he was already famous, not for anything he or his older brother or even his father had achieved, but because he was the grandson of Erasmus Darwin, a man with a European-wide intellectual, medical, scientific, and literary reputation. Erasmus was also remembered for his radicalism; he had been an anti-clerical, anti-monarchical Whig, supporter of both the American and French Revolutions, a denier of the existence of the human soul, and a proponent of the theory of evolution.

Erasmus Darwin was as famous—indeed, *more* famous—than the other great evolutionist of the time, the Frenchman Jean Baptiste Lamarck, and it was probably Erasmus Darwin's account of the mechanics of evolution in *Zoönomia* that influenced Lamarck's presentation of transmutationism in his *Philosophie Zoologique* published almost fifteen years later in 1809, the very year of Charles Darwin's birth.[13] Erasmus's fame reached its peak at the turn of the eighteenth into the nineteenth century.

When Charles arrived at Edinburgh in October 1825, all eyes were on him, and at first he was excited to be there. He had escaped the asphyxiating degradation of Shrewsbury School and was living as a young man on his own, yet with all the comfort of being in close company with his dear brother Erasmus. Erasmus, however, was struggling. He was up to the mark intellectually, but his health, never strong, seemed to be getting worse under the pressures of medical school.

Robert had high hopes for his sons, but also deep reasons to worry. Both his sons had been named after dead brothers. Erasmus had been born on December 29, 1804, exactly five years after his namesake, Robert's brother, had committed suicide over business debts, throwing himself into a river. With that background, Robert was not going to push Erasmus too hard; and if Erasmus became too frail to practice medicine, Charles would have to carry on the tradition.

By Charles's second year at medical school (1826–27) Erasmus had gone on for more study in London, and his place as friend and confidant was taken by Robert Grant. In his autobiography, Darwin notes, almost as an aside, that "one day, when we were walking together" Grant

> burst forth in high admiration of Lamarck and his views
> on evolution. I listened in silent astonishment, and as far as I

can judge, without any effect on my mind. I had previously read the *Zoönomia* of my grandfather, in which similar views are maintained, but without producing any effect on me. Nevertheless it is probable that the hearing rather early in life such views maintained and praised may have favoured my upholding them under a different form in my *Origin of Species*.[14]

The notion one might get from this episode is of Darwin being suddenly bumped by a wild theory incongruously spouted by Grant. But in fact Grant had sought out Charles Darwin at Edinburgh precisely *because* he was the grandson of Erasmus Darwin the transmutationist, and it was Erasmus, so Grant himself explained, that "first opened my mind to some of the laws of organic life."[15]

The impression given by Charles, that Grant was more like a passing acquaintance than an intimate friend, is quite misleading. During his second year at Edinburgh, Darwin actually became extremely close to Grant. When Charles was supposed to be working hard at his medical studies, he was instead working diligently under Grant for several months like a devoted disciple as he pursued his research on polyps. Although he never said so directly, it is reasonable to assume that Charles gave short shrift to his relationship with Grant because he was bitter that Grant had co-opted his research without attribution.[16]

The goal of this research was directly tied to Grant's desire to demonstrate that transmutationism was correct. He hoped to show, from the minute study of polyps, that the dividing line between plants and animals was not a line at all, but a continuous evolutionary smear. That is *exactly* the strategy Darwin would take in later life in his exhausting and tedious investigation into everything from barnacles and orchids to earthworms. It was Grant who first taught Darwin to look at the details of nature through Erasmus Darwin's eager eyes. Though Darwin had already studied his grandfather's *Zoönomia* and read the French evolutionist Lamarck, including his well-known lecture on species transmutation,[17] it was Grant who brought it to life and it was Grant who showed Darwin what transmutationist research should look like. Erasmus Darwin had provided the speculative framework (including ideas that Charles would make famous, such as common descent with modification, sexual selection, the survival of the fittest); it was transmutationist research that could provide the evidence.

So Darwin was not entirely idle at medical school. Certainly when it came to his medical studies he was a slacker; but when it came to the *other* family tradition—developing the theory of evolution—Darwin was as eager a pupil as any. His grandfather Erasmus Darwin had even added to the family's coat of arms an evolutionary decoration: three scallop shells and the motto *E Conchis Omnia,* "All things out of shells," a secretive shorthand capturing in terse Latin his belief that all life evolved from a common ancestor.[18] Charles's father Robert adopted *E Conchis Omnia* in the 1790s as his own motto, displaying it on his bookplate.[19] The great difference between Erasmus and Robert was this: Robert kept his views entirely private. He had no wish for the notoriety as a radical that his father had. And Grant was exactly the sort of person Robert would not have wanted Charles to befriend—an explicit evolutionist whose larger radical, social-political vision was the antithesis of the social respectability that Robert Darwin jealously defended for his family.

Robert wanted Charles to succeed at medical school, but Charles was too much his father's son in at least one regard: he really had no inclination toward medicine to begin with, and soon enough, he had a positive disinclination because he couldn't stand the sight of blood. "I . . . attended on two occasions the operating theatre in the hospital at Edinburgh, and saw two very bad operations, one on a child, but I rushed away before they were completed."[20]

By the summer of 1826, he knew he wasn't meant for medicine, but the difficult task remained: how to tell his father? That was a difficult task precisely *because* Robert Darwin himself had no love of medicine—so that was no excuse. Robert had manfully forced himself to overcome his distaste for medicine and his horror at the sight of blood. So could Charles.

Moreover, Charles knew that he didn't need to find a profession at all. Not only did his father's practice ensure that the family was very well off, but shrewd capital investments had actually made Robert Darwin quite rich. Charles could, if he chose and his father allowed it, be as idle as any gentleman's son. "I never imagined that I should be so rich a man as I am," wrote Darwin, struck by the immensity of his expected inheritance, and this "belief was sufficient to check any strenuous effort to learn medicine."[21]

The effort saved from intensive study was better spent during his time at Edinburgh riding, shooting, reading novels, eating good food with good friends, and collecting rock, insect, and animal specimens.

But a day of reckoning arrived. "To my deep mortification my father once said to me, 'You care for nothing but shooting, dogs, and rat-catching, and you will be a disgrace to yourself and all your family.'"[22] As Charles conceded, "He was very properly vehement against my turning an idle sporting man, which then seemed my probable destination."[23]

Robert's words stung. Charles's father was a keen and demanding man, "generally in high spirits," laughing and joking even with his servants, "yet he had the art of making every one obey him to the letter. Many persons were much afraid of him." He was both loving and prickly, "very sensitive so that many small events annoyed or pained him much" and "easily made very angry, but as his kindness was unbounded, he was widely and deeply loved."[24] But "the most remarkable power which my father possessed was that of reading the characters, and even the thoughts of those whom he saw even for a short time."[25]

Or for a long time, as with his own sons. Charles had actually gone on rounds with his father before entering Edinburgh, and that gave Robert a chance to judge his capacities and character. "My father, who was by far the best judge of character whom I ever knew, declared that I should make a successful physician—meaning by this, one who got many patients."[26] Robert had judged his son a worthy successor, yet Charles had failed him. If Charles considered his father "the best judge of character whom I ever knew," how heavily and prophetically those words must have fallen upon him: "You will be a disgrace to yourself and all your family."

But what to do with a son unfit for medicine? Robert proposed that Charles "should become a clergyman"[27]—a proposal that tells us as much about the state of the Anglican Church at the time as it does about Charles and Robert Darwin, for if there was one sure inheritance that had passed through three generations of Darwins, it was religious skepticism.

The Deism of grandfather Erasmus was not a faded form of Christianity; it was not monotheism with all the Trinitarianism rubbed off by reason. Enlightenment Deism, such as Erasmus embraced, was based on radical skepticism about the Bible, Christian revelation, and Christianity in general. Erasmus's Deism was enough to shock even a Unitarian like Samuel Taylor Coleridge. Coleridge visited the renowned poet-physician-philosopher in early 1796. "He had heard that I was a Unitarian," reported Coleridge, "and bantered incessantly on the subject of Religion." Erasmus had evidently gone far beyond the denial of the Holy Trinity, and seemingly, a good way past the lightly laid boundaries of Unitarianism, the settling-in Church of Deism. "He is an Atheist," declared Coleridge, "but he has no new arguments. . . . When he talks on any other subject he is a wonderfully entertaining and instructive man."[28] Erasmus famously described Unitarianism as a featherbed to catch a falling Christian. Whatever Erasmus was, he was beyond that.

Charles Darwin's father Robert needed no featherbed, as he seems to have been an atheist,[29] but an atheist who was both circumspect in expressing his views and who supported the Anglican Church as a buttress against the barbarism of the lower orders who, unless tamed and restrained by religion, would recreate in England the horrors of the French Revolution. Radical thought, while fine enough if it circulated quietly among the upper, closed circles of society, was too heady a wine for the masses—or for women, who, as weaker vessels, also needed the crutch of religion, he believed.

Charles's own religious "training" was in Unitarianism under the tutelage of his sisters. Unitarianism had begun as a kind of biblically-based denial of the Trinity in the early Reformation, but by Darwin's time it had become the church for the smart-set, who were smugly certain that the Bible was merely one more book of ancient mythology. Religion was certainly not as high a calling as medicine, but it was socially respectable, and, in Robert's view, a necessary support for public order. Certainly Charles would have graduated from his sister's Unitatianism to his father's skepticism by the time he was a young man.

So it was in all good British conscience that Robert Darwin could propose, and Charles accept, the notion that a doubting Whig could matriculate in a Tory religious institution, and take up the life of a country parson. There, he could continue with his "shooting, dogs, and rat-catching," and not be "a disgrace" to himself and his family.

Yet, however much this relieved Charles, the notion of his being an Anglican clergyman had its lackluster side. His "vocation" to the church represented a safe haven for a Darwin not up to the grade. He knew that he was taking a path overshadowed by disappointment. The prospect of ordination and a life of genteel hypocrisy carried no esteem from his father, for whom Charles would thereafter be something of a lovable loser. Worse than that, his brother Erasmus, the brightest third generation skeptic, would have nothing but scorn for Charles's profession. This was by far the worst blow, as Charles valued his brother's opinion even more than that of his father. Charles Darwin may not have wanted to be a doctor of medicine, but being a doctor of divinity seemed to be no step up—though he would soon find reason to change his mind.

Chapter 2

Theology School

When his father set before him the prospect of becoming an Anglican priest, Charles asked for time to think it over, being somewhat worried about having to declare his allegiance to "all the dogmas of the Church of England," especially since, following upon two generations of freethinkers, he wasn't all that familiar with them. So he read a few theology books, and "as I did not then in the least doubt the strict and literal truth of every word in the Bible, I soon persuaded myself that our Creed must be fully accepted."[1]

Given his background, this statement rather stretches credulity. It is found in his autobiography—written much later, and which he intended to be read by his family, not for general publication—and more likely its purpose is as a rhetorical device: he wanted to present a didactic tale where his religious skepticism was the result of scientific discovery rather than an inheritance; it also grounded his life story in an already well-ploughed, liberal, Whig historical scheme where reason triumphs over superstition, science over religion, industry over piety. But, as we'll see, that's not really how Darwin's ideas developed.

After a little tutoring to bone up on his Classics, Charles arrived at Christ's College, Cambridge, in January 1828, to study for the only degree he ever earned: a bachelor of arts degree that was essentially a pre-divinity degree, its purpose being to offer the student a well-rounded liberal arts education. Cambridge was a place of privilege, a bastion of conservatism, defined by a tightly bound union of Tory politics and Anglican religion. The curriculum asked very little of its students, and that is just what Darwin was prepared to give it. It was a place for young gentlemen, and like many a gentleman's son—and also, we might add, many an Anglican parson—Darwin threw himself into natural history, the catch-all term of the time meaning anything from bug-collecting to geology, or as his father called it, shooting, dogs, and rat-catching. Here at Cambridge, he could fulfill his father's low estimation of his character.

Darwin was still a homebody, never really comfortable unless he was near family. Happily, he discovered some very amiable cousins from the Wedgwood-Darwin family alliance, the cousins Hensleigh Wedgwood and William Darwin Fox. In particular, Fox became a substitute brother, but one who, unlike Erasmus, was just as keen as Charles was on foxhunting and collecting rocks, worms, and insects. Fox's rooms were filled with stuffed swans and martens, a march of caterpillars, pinned specimens of moths and butterflies, and curious plants and every kind of natural whatnot—an ever burgeoning mini-museum that was the despair of his servant charged with keeping his rooms tidy.

Fox, the sportsman and collector, was the perfect soul mate for Darwin, who was himself a crack shot and adept horseman, and who adored not only hunting for rabbits, foxes, and partridges, but for exotic beetles. No "pursuit at Cambridge was followed with nearly so much eagerness or gave me so much pleasure as collecting beetles," he wrote, and "no poet ever felt more delight at seeing his first poem published than I did at seeing in Stephen's *Illustrations of British Insects* the magic words, 'captured by C. Darwin, Esq.'"[2]

His appetite for collecting, for getting identities exactly right, for accurate categorization and naming was voracious and admirable—and unlike his cousin William Fox, this was not a mere gentleman's hobby for Darwin. It was a vocation, the very kind of thing he knew he was made for. He was no doctor, poet, or philosopher, nor was he a money-maker. He was a naturalist—and luckily, many parsons of the nineteenth century were naturalists themselves. Fox, in fact, became one: a happy country priest with few religious duties, a good living, and lots of time for sport—exactly the future Robert intended for Charles, and one that Charles could easily have gravitated to.

Good living and sport, if not religion, were much to Darwin's taste, even if he put his own stamp on these gentlemanly pursuits. While it was common for a gentleman to enjoy a stout breakfast, a robust day's hunting, and a sociable supper laid out by the servants, Darwin and his companions formed a Glutton Club that, in contrast to the other dining societies, aimed at consuming "birds & beasts which were before unknown to human palate."[3] Hawk, bittern, and owl were among the items on the menu, and given that Charles had once popped a beetle in his mouth so he could free a hand to catch another, it seems fair to guess that his palate ranged far afield.

He also made a great show of enjoying music, even hiring singers to perform in his rooms, but had to confess that "I am so utterly destitute of an ear, that I cannot perceive a discord, or keep time and hum a tune correctly." He was even unable to identify a common song (such as "God save the King") if it was played faster or slower than usual or on an unaccustomed instrument.[4]

Meanwhile, Darwin flirted with one Fanny Owen back at his home county. She was, in his estimation, "the prettiest, plumpest charming Personage that Shropshire possesses."[5] She also defied convention and showed her astonished, enchanted, and firearms-loving suitor that she knew how to shoot a gun. But nature provided a greater romantic attraction. Darwin became a disciple of John Henslow, a professor of botany, whose classes were part of his undergraduate curriculum. At Cambridge, natural history was taken to be an essential element of natural theology. As Darwin recalled of Henslow, "He was deeply religious, and so orthodox, that he told me one day, he should be grieved if a single word of the Thirty-nine Articles were altered. His moral qualities were in every way admirable. He was free from every tinge of vanity or other petty feeling; and I never saw a man who thought so little about himself or his own concerns."[6]

In Henslow, an ordained Anglican priest, Darwin saw, perhaps for the first time, a sincere and deep union of the love of nature and the love of God. Here was a gracious man in every way; a man who was kind and who gloried in every aspect of

God's creation. Moreover, Henslow praised Darwin's intellect, he encouraged his naturalist's avocation, and so he provided the sort of affirmation Darwin had not received from his father. Henslow did not regard Charles as a lovable failure, but as a gifted pupil, and Darwin responded with enthusiasm.

Another scientist-priest who took Darwin under his wing was Adam Sedgwick, an Anglican clergyman who was one of the founders of modern geology. Without Sedgwick, Darwin would likely never have become a competent geologist, and geology was an absolutely essential underpinning to his theory—or any theory—of evolution. I think it is fair to say that, under the guidance of both Henslow and Sedgwick (and with the friendship of William Fox) Charles became better reconciled to the idea of becoming an Anglican priest. This does not mean that he became a scriptural literalist—neither Henslow nor Sedgwick were literalists—but it does show that Darwin could easily have become a man like the Anglican stalwart, and natural theologian, William Paley, author of *Principles of Moral and Political Philosophy* (1785), *Evidences of Christianity* (1794), and *Natural Theology* (1802). Darwin read Paley in preparation for his bachelor's degree examinations. Like Henslow and Sedgwick, Paley argued that nature declared the glory of God in every detail. Had Darwin followed down this path and pursued the more detailed theological training that would have led to his ordination, had he become a country parson propounding a theory of God-guided evolution, it is possible that the entire intellectual history of the West might have turned out differently.

Ironically, it was Henslow who diverted him from the further pursuit of theology. Henlsow was a Romantic, with a capital

"R," and, as with Darwin, his greatest romance was with nature. Nature was a wild, pulsating, extravagant bazaar of bursting novelty, of exotic dramas that needed to be seen and experienced, not merely read about—though Henslow and Darwin were devoted readers of nature-focused travel books, and were especially influenced by the heady German Romantic, Alexander von Humboldt and his *Personal Narrative* detailing his trek in 1799–1804 through the Brazilian rain forest.

Henslow longed for adventure, and passed that longing on to Darwin. For the first time Darwin, the home-loving Englishman, passionately desired to be transported to strange lands, to document rarities that could not be found in England. It was through Henslow's machinations that Darwin received a surprising letter on August 29, 1831, inviting him aboard the HMS *Beagle* as a gentlemanly companion (not, as Darwin later claimed, as the ship's official naturalist, a job that went, by tradition, to the ship's physician) to the aristocratic Captain Robert FitzRoy. Henslow, a family man, would travel vicariously through his disciple. The ship was slated to leave in four weeks. It would be the great adventure of Darwin's life.

Chapter 3

The Great Adventurer

In accepting a spot aboard the *Beagle*, Darwin agreed to sail around the world on a ninety-foot boat riddled with sea rot, crammed with seventy-four people, and captained by a twenty-six-year-old Tory aristocrat. Darwin, the great homebody, was about as unlikely a candidate for such an adventure as one could find.

Certainly his father thought so. Charles told Henslow that "my Father, although he does not decidedly refuse me, gives such strong advice against going . . . that I should not be comfortable, if I did not follow it." He added, "If it had not been for my Father, I would have taken all risks. . . . Even if I was to go, my Father disliking would take away all energy. . . . "[1] As far as Robert was concerned, the trip promised to be a dangerous and unprofitable distraction from Charles's second-rate, but still socially respectable vocation as an Anglican vicar. Charles, on the other hand, saw the adventure as an opportunity to advance his true vocation as a gentleman naturalist. But here was the rub: the position of Captain's companion was unpaid, so he needed his father's financial support. Robert gave Charles a glimmer of hope when he proposed that if "any man of common sense" viewed the trip as anything other than folly, then he would allow his son to go.

The next day, while off shooting at the Wedgwood estate, Charles set to scheming with his Uncle Josiah Wedgwood. Uncle Jos agreed to be the man of "common sense." Armed with a well-honed list of answers to Robert's reservations, Uncle Jos returned with Charles to the Darwin home in Shrewsbury and told Robert that the prospect of his son joining the crew of the *Beagle* was a splendid idea. To Charles's delight, his father "at once consented in the kindest manner." Much overcome, Charles offered a conciliatory half-promise, appealing to the doctor's characteristically tight hold on the finances, that he'd have to be "deuced clever" indeed if he managed to spend more on ship than he did each year at Cambridge. Robert "answered with a smile, 'But they all tell me you are very clever.'"[2] This affectionate, offhand recognition of his son's intelligence, remembered with gratitude so many years later, was a great healing salve to Charles. Failing at medicine, and shuffled off with a sigh to become a parson, Charles could now prove himself worthy of the Darwin name as a clever naturalist (and that included spending much more of his father's money on his trip round the world than he had at Cambridge).

On September 1, 1831, Darwin gleefully accepted Captain FitzRoy's offer, "as happy as a king." "Woe unto ye beetles of

South America!" he vowed triumphantly.[3] Alas, it would be some time until South American beetles felt the weight of this warning, because it took much longer than expected to get the *Beagle* up to snuff.

During this period of delay, Darwin's moods fluctuated. At times he was lighthearted, almost giddy. Writing in mid-November to one of his Cambridge friends, C. T. Whitley, Darwin told him that he'd have come down to London for a "day of victory & triumph & inward-glorying" with the Glutton Club if he'd known how long the delay was going to be. He eagerly described his pending trip, pouring out his pent-up excitement in writing: "The scheme is a most magnificent one. We spend about 2 years in S America, the rest of time larking round the world." Darwin promised to beat the best of them in "telling lies" when he came back.[4]

At other times, the delay almost wore down his resolve to go on the trip at all, especially since the anxiety caused him heart palpitations. His fretfulness also gave way to spasms of self-doubt. Perhaps he couldn't do it. Perhaps his father was right after all. He wasn't fit to be a sailor. He struggled with these gloomy thoughts as the period of waiting wore on.

During this time he received a letter from the Reverend Henslow, which offered pastoral advice for a young gentleman—even a member of the Glutton Club—unused to the coarse atmosphere of a ship. Darwin must not expect to find on the Beagle the sort of gentility he had known all his life. He must not be too quick to take offense "at rudeness of manners & any thing bordering upon ungentlemanlike behavior." This was not as a mere matter of prudence, but of Christian forbearance, of working after spiritual perfection. Henslow warned that Darwin would "infallibly be subjected to...coarse and vulgar behavior" among his comrades over the course of the expedition. "Take St James's advice & bridle your tongue,"[5] he urged his charge. As excited as he was about Charles's opportunity to see the world and discover more about nature, the Reverend Henslow cared more about how Charles fared on his spiritual journey.

Finally, after excruciating delays, the *Beagle* sailed from Plymouth, England, on December 27, 1831. They embarked from the same port from which Sir John Hawkins had begun England's Atlantic slave trade in the sixteenth century, and which in the century after had been the launch of Pilgrims sailing for America. In retrospect, Darwin understood that "the voyage of the *Beagle* has been by far the most important event in my life and has determined my whole career." The beginning of this pivotal point in his life was marked, he later noted, by "so small a circumstance as my uncle offering to drive me 30 miles to Shrewsbury, which few uncles would have done," to talk to his father about letting Charles sail aboard the Beagle, "and on such a trifle as the shape of my nose," "the weaknesses of which Captain FitzRoy was willing to overlook."

Now the adventure was begun in earnest. The *Beagle* was packed full of cargo—not just with the usual store of provisions, but with an unusual number of scientific instruments. The *Beagle*'s primary mission was to survey the coastline of South America, especially its southeastern side. But it was also meant as a voyage of exploration and discovery, which would allow Darwin plentiful landfalls in which to travel into the South American interior and collect specimens. Even more intriguing, Captain FitzRoy was aiming to return three rather interesting specimens from the *Beagle*'s first South American survey (1826–1830). These were human beings, natives of Tierra del Fuego, named Yokcushlu, Orundellico, and El'leparu (better known by the odd English names they were given, Fuegia Basket, Jemmy Button, and York Minster). A fourth, Boat Memory, had died of smallpox in England. The three survivors had been Christianized and were returning home as Gospel salt for their savage tribes.

Despite his enthusiasm, Charles almost immediately confirmed his father's doubts about his seaworthiness. He became violently seasick. In fact, he was wretchedly seasick for the entire voyage, except of course in the blessed times when he was exploring on land. Happily for Darwin, of the nearly five years he spent as a member of the crew of the *Beagle*, two-thirds of his accumulated time was spent exploring *terra firma*. The sea, however, became a subject of dread. "I loathe, I abhor the sea and all ships which sail on it," Darwin later wrote home in queasy self-pity. "Not even the thrill of geology makes up for the misery and vexation of spirit that comes with sea-sickness."[6]

FitzRoy was the very picture of sympathy, as were the rest of the crew. The Fuegian Jemmy Button would sidle up to Darwin, who was green-gilled and lying in a hammock that pitched mercilessly with the ocean swells, and offer amused sympathy, "Poor, poor fellow!"[7] But violent nausea wouldn't end when he finally stepped off ship. The voyage of the *Beagle* marked the beginning of Darwin's life-long struggle against his stomach. Whatever the cause of his perpetual bouts of retching later on—

a strange "bug" picked up on his odyssey, frail nerves, his addiction to taking snuff, a diet rich in sweets, a hereditary malady—he spent nearly his entire life as if he'd never gotten off the *Beagle*, suffering long periods of debilitating nausea and vomiting, accompanied by headaches, interrupted only occasionally by bouts of good health. Whatever our romantic notions of Darwin the scientist, his work during and after the *Beagle* was nearly always carried on as a stumbling man on a wambling deck—despite which he maintained surprisingly good spirits.

FitzRoy and Darwin, Tory and Whig, became tight-knit right away. They dined together, an especial privilege that showed where Darwin stood on the ship's pecking order. Aside from a few episodic sparks cast up by raw nerves and the Captain's somewhat mercurial temper—and despite the friction that developed after the voyage—they were fast and congenial friends on board the *Beagle*. It was not just that they came from the same upper reaches of English society. FitzRoy was quite sharp, with a scientific mind and a passion for exact measurement, careful observation, and detailed scientific journaling that provided a model for Darwin.

Darwin was first able to set his feet on dry land again in mid-January at St. James (Santiago, or St. Jago, as Darwin called it), the largest of the Cape Verde islands off the west coast of Africa. He was transported into a kind of intellectual, aesthetic ecstasy. "Here I first saw the glory of tropical vegetation," Darwin wrote. The Romantic Humboldt had conjured a glorious vision of exotic nature, so glorious that Darwin feared it must surely be exaggerated. But upon exploring St. James, "how utterly vain such a fear is." Mere human words, even the most poetically inspired, fell far short of the actual experience. "It has been for me a glorious day, like giving to a blind man eyes—he is overwhelmed with what he sees & cannot justly comprehend it."[8] The real vibrant beauty of the island's flora and fauna, so different from England, was a revelation, far exceeding mere human words and expectations. The voyage had only just begun; and for Darwin, it could not, sea-sickness aside, have begun better.

Years later, when he had turned the scribblings of his shipboard journal into his first major literary effort, *Journal of Researches into the Geology and Natural History of the Various Countries Visited by H. M. S.* Beagle, *under the Command of Captain FitzRoy, R.N. from 1832 to 1836*, he muffled his romantic enthusiasm in more staid English understatement. Yet the original youthful wonder bubbles through the tight Victorian collar. Viewed from the sea, Darwin tells the reader, the port at St. Jago "wears a desolate aspect." Yet, the "scene . . . is one of interest; if, indeed, a person, fresh from the sea, and who has just walked, for the first time, in a grove of cocoa-nut trees, can be a judge of any thing but his own happiness."[9] Euphoria was more like it; Darwin was overflowing with wonder, a real scientist in his natural habitat. As the Greek philosopher Aristotle noted over two millennia before, it "is because of wondering that men began to philosophize and do so now."[10] On the *Beagle*, Darwin earned the nickname "Philos," short for natural philosopher, the nineteenth century moniker of a "scientist."

While on St. James Island he galloped off with some of his crewmates to visit an old wreck of a town, Ribeira Grande, which had once been an immensely rich slave-trading center for the Portuguese, but which now lay in ruins. Darwin secured a "black padre" as his guide, and when it was time to leave, "We presented the black priest with a few shillings, and the Spaniard [translator] patting him on the head said with much candour, he thought his colour made no great difference." [11]

The Darwin family had a tradition of opposing slavery—and indeed, this became an item of temporary conflict between Darwin, who opposed slavery, and Captain FitzRoy, who defended it. Two generations before, Erasmus and an earlier Josiah Wedgwood (Uncle Jos's father) had thrown their lot in with the great humanitarian and Tory politician William Wilberforce in the long, slow effort to extinguish slavery in the British Empire, two liberal skeptics yoked to a fervent conservative evangelical Christian for a most noble cause. Britain had abolished the slave trade in 1807, and enforced that abolition via the Royal Navy. But it took another quarter century to abolish slavery itself within the empire (it was practiced chiefly in the British West Indies). In 1833, the year after Darwin viewed the ruins of Ribeira Grande, Britain passed the Slavery Abolition Act, Wilberforce expiring three days after hearing of its assured passage, while the *Beagle* sailed down the east coast of South America from Maldonado to the Rio Negro.

In his *Journal*, Darwin recorded no indignant remarks about the history of Ribeira Grande. But his next exposure to slavery was different. In April 1832, on a landward journey from Rio de Janeiro, he met the master of an estate on the Rio Macae, an expatriate Irishman named Patrick Lennon. He witnessed Lennon, after a quarrel with the foreman, on the verge of separating slave families and selling the women and children at auction. "Interest, and not any feeling of compassion, prevented this act. Indeed, I do not believe the inhumanity of separating thirty families, who had lived together for many years, even occurred to the person." It horrified Darwin, who then reported an episode that struck him "more forcibly than any story of cruelty."

> I was crossing a ferry with a negro, who was uncommonly stupid. In endeavoring to make him understand, I talked loud, and made signs, in doing which I passed my hand near his face. He, I suppose, thought I was in a passion, and was going to strike him; for instantly, with a frightened look and half-shut eyes, he dropped his hands. I shall never forget my feelings of surprise, disgust, and shame, at seeing a great powerful man afraid even to ward off a blow, directed, as he thought, at his face. This man had been trained to a degradation lower than the slavery of the most helpless animal.[12]

In reading this passage, one can feel the heat from Darwin's deep moral indignation even now. Yet perhaps even more important is how this indignation sat increasingly uneasily with the thoughts of Darwin the scientist who would not only categorize races, but place them in an evolutionary order of inferior and superior, with black and white being practically different species. Some of these races, Darwin would maintain, were destined for evolution by extinction.

But we are getting ahead of ourselves on the journey, for another important incident occurred on his first landfall, St. James. He noted a curious fact on this volcanic island. Up on the cliffs, far beyond the conceivable reach of the tides, he could see a white horizontal band of rock. He inspected it and discovered that it contained sea shells of the very same type that were found on the beach. Darwin certainly wasn't the first to notice the white ribbon, but noticing and wondering, and analyzing and explaining are different things.

Darwin did not come to the mysterious white line unarmed. He'd already done a significant amount of geological work with Sedgwick, and even more important, FitzRoy had made Darwin a gift of the first volume of Charles Lyell's *Principles of Geology*, published the year before the *Beagle* set to sea. "The very first place which I examined, namely St. Jago in the Cape Verde islands, showed me clearly the wonderful superiority of Lyell's manner of treating geology, compared with that of any other author, whose works I had with me or ever afterwards read."[13]

Interestingly, among Lyell's chief rivals and disputants was Darwin's mentor Adam Sedgwick (with Darwin's other Cambridge mentor, John Henslow, on Sedgwick's side). Sedgwick and Lyell are considered the founding fathers of modern geology. Both were anti-transmutationists, adamantly opposed to the sort of evolutionary theories espoused by Lamarck (and Erasmus Darwin), and neither one was a scriptural fundamentalist, though Sedgwick and Henslow were Anglican priests (Lyell was a Deist).

What set them apart as geologists is that Sedgwick argued that all the evidence of the fossil record showed that it was progressive, moving from an abrupt, well-defined beginning in the Cambrian period through definite stages of more complex creatures adapted to the developing Earth, and culminating in the one creature capable of being a scientist, man. For him, it was a pattern established in Genesis but written on a much grander scale of time.

Lyell would have none of that. Scripture and geology must be kept entirely separate, and so Lyell set up geology as a science defined to keep them that way. He conjured up a geological vision of slow undulations of the earth, stretching from eternity to eternity, entirely indifferent to the biblical historical scheme. In short, he believed that, appearances to the contrary, there was no march of progress in the fossil record, only an endless wandering. While the fossil record looked progressive, future discoveries would reveal that there was really no direction at all, just an ebb and flow of species that matched the aimless geological changes. These changes were not arbitrary or unknowable. All changes in geology came about through the forces we see around us today, volcanic uplift, rain, erosion; everything must be the result of a relentless process guided by eternal laws. But since the laws were eternal and the geological processes therefore so regular, geology gave no clue of anything extraordinary, anything singular that would mark a beginning of time or an end (even if there were a definite beginning).

Darwin therefore set foot on St. James's beach with two theories, two visions. Even with his great love and respect for Sedgwick and Henslow, Charles chose Lyell as his guide, and became a deeply devoted partisan. He set about providing a Lyellian interpretation of that strange white band above his head, one that delighted him immensely and would much impress the eminent geologist once he'd heard of it. Darwin posited that, in accord with Lyell's ideas, that white band of shells was the result of constant geological shifts on the island, that the cliffs were once the domain of the sea, but had been pushed upward by slow geological forces working from underneath. Lyell believed in a doctrine called uniformitarianism that rejected dramatic beginnings and endings, things that might evoke the idea of creation. He insisted on gradual and constant geological transformations. Darwin was very taken with this idea, and it is certain that the notion of slow transformation influenced his own later theories of descent with modification, despite Lyell's own doubts about evolution.

Leaving Cape Verde, the *Beagle* set off across the Atlantic, aiming for the eastern shores of South America, stopping first at Bahia, then Rio de Janeiro, Montevideo, and finally, by December 1832, at Tierra del Fuego. No one, certainly not a homebody like Darwin, could have asked for more adventure. In August 1832 he had two whiffs of near combat. At the harbor in Buenos Aires, a Spanish ship fired across the bow of the *Beagle*. Allegedly it was a warning shot—that the city was under quarantine—but FitzRoy assumed it was an act of perfidious aggression. He ordered his cannons loaded, ready to retaliate, and sailed past the Spanish ship, shouting that if it dared shoot again, "we shall send our whole broadside into your rotten hulk."[14] Charles was ready for action, hoping to see FitzRoy riddle their ships, but was left disappointed.

But if not an amateur Hornblower or Nelson, perhaps he could be an amateur Wellington or Sharpe. At Montevideo, east of Buenos Aires, across the River Plate, mutinous black soldiers had taken over the central fort in town. Montevideo's police chief begged FitzRoy to help restore order. With British residents in danger, the captain knew his duty. He led half a hundred of the ship's men, fully armed, marching on the fort. Darwin, armed with pistols and sword, brought up the rear, hoping for action. But the rebels gave up without a shot, much to Darwin's disappointment. "There certainly is a great deal of pleasure in the excitement of this sort of work," he mused,[15] and perhaps Darwin the hunter, the shooter, and now the adventurer might easily have had a vocation as an officer in the British Army, especially in this, Britain's imperial phase, when he could have lived in all the most exotic locales and examined all the most exotic species from India to Africa to the Caribbean, and even helped fight against the slave trade. But however agreeable he found the work, at least in its temporary excitement, it was not, he confessed, the sort of work a philosopher like himself expected to be called upon to perform.

No, Darwin knew that his greatest pleasures and talents were those of the naturalist. In his forays on land, he inspected every item of flora and fauna with wonder, collected as many important specimens as he could lay hands on, and packed them off to Henslow back in England. His gentleman's ability to ride and shoot were put to good use, whether for bagging specimens or bagging dinner.

Continually immersed in exoticism, he was not entirely cut off from England. He could send and receive mail, which turned out at times to be a mixed blessing. On the way down the South American coast, he had received in early Spring of 1832 a letter from home informing him that his first romantic flame, Fanny Owen, had married. He was certainly not the first sailor to get a "Dear John" or "Dear Charles" letter. He tried to put a brave face on it, but it pained him just the same.

He received a much happier package in late 1832, a yearned-for copy of the second volume of Lyell's *Principles of Geology*, which opened with a presentation and then refutation of Lamarck's transmutationism. Reading Lyell's attack must have produced a conflicting array of feelings, especially since the geological argument of Lyell would seem to fit so nicely with the aimless, endless undulations of species transforming one into another. But Lyell would have none of it. After setting out the "machinery of the Lamarckian system," Lyell included a several page exhibition of the theory (in a sarcastic tone) that focused on the alleged transformation of an "orang-outang, having been already evolved out of a monad . . . made slowly to attain the attributes and dignity of man."[16] After sketching this alleged Lamarckian absurdity, Lyell provided a detailed presentation of the defects of transmutationism. First, Lyell argued, Lamarck had confused limited variation *within* a species, which does occur, with unlimited variation which could produce radically *different* species, for which there is no evidence.[17] Second and related, we find that species for which we have detailed anatomical evidence over long periods of time have not changed significantly but only varied, if at all, within the limits of the species. Lyell used the example of ancient embalmed cats from Egypt and present day cats which exist under a myriad of conditions worldwide.[18] They were all basically the same. Moreover, Lyell pointed out, domestic breeders find that, no matter how hard they try, they cannot, through breeding, transform the animals beyond certain innate boundaries.[19] In fact, the amount of change breeders can bring about diminishes with each generation (rather than increasing, or at least remaining constant, which transmutationism demands).[20] Furthermore, the traits that are bred do not arise *ex nihilo*, but are implicit to the species and are brought to the surface through breeding (an interesting anticipation of the place of genes in DNA). They do not signal a move *toward* a new species, but an expression of something already *within* the defined species. Lyell offers an interesting example: the elephant which, taken from the wild, expresses extraordinary intelligence through human training in a very short time, much more intelligence than "the orang-out-ang" that, merely because of physical resemblance, Lamarck took to be a proto-human.[21]

Obviously the notion that Darwin wasn't really pondering evolution as a theory on his voyage until he knocked into the Galapagos Islands on the other side of South America is unsupportable. Before the *Beagle* reached Tierra del Fuego, his newfound adulation of Lyell threw him right into the midst of the controversy, and in a very peculiar way. Lyell was taking the cudgels to the kind of transmutationism championed by his grandfather and quietly endorsed by his father.

But it was as a student, however reluctant, of theology that Darwin received his next *Beagle*-borne lesson. Captain FitzRoy, despite being a defender of slavery, endorsed the wildly counterintuitive Christian view that human beings, no matter how strange and savage their appearance, are all made in the image of God. With that belief in hand, FitzRoy was bent on the conversion of his brothers and sisters, the Fuegians.

Darwin didn't need Christianity to declare universal brotherhood; that was part of his Enlightenment heritage. But that was largely a party opinion untried by the experience of meeting actual natives, and Darwin soon struggled with the conflict between experience and ideology.

When they sailed into Tierra del Fuego, "A group of Fuegians partly concealed by the entangled forest, were perched on a wild point overhanging the sea; and as we passed by, they sprang up, and waving their tattered cloaks sent forth a loud and sonorous shout. The savages followed the ship, and just before dark we saw their fire, and again heard their wild cry."[22] The next day, Darwin met them face to face, a tall people, powerful, "skin of a dirty coppery red colour," barely clad, and hair "black, coarse, and entangled," with red and white stripes painted across their faces, "closely resembling the devils which come on the stage" in plays Darwin had seen. "I could not have believed how wide was the difference, between savage and civilized man. It is greater than between a wild and domesticated animal, in as much as in man there is a greater power of improvement."[23]

Here, he was obviously thinking not only of the difference between an Englishman and a Fuegian, but perhaps more of Fuegia Basket, Jemmy Button, and York Minster, decked out in the accoutrements of their short immersion into Anglicanism. For these Fuegians transformed into English civility, the meeting proved uncomfortable, even insulting. They were repulsed, insisting that these particular ill-clad Fuegians were from a different and second-rate tribe. Indeed, they were insulted, as if a captured English gentleman were set down on the shore of Ireland and expected to speak congenially and as an equal with an Irish peasant. As FitzRoy reported, Jemmy and York took them to be "monkeys—dirty—fools—not men."[24]

And *these* savages, Darwin remarked, were even of the better sort. They were "a very different race from the stunted miserable wretches further to the westward," the "most abject and miserable creatures I any where beheld. . . . Viewing such men, one can hardly make oneself believe they are fellow-creatures, and inhabitants of the same world."[25]

Later in life, Darwin's experience with these savage tribes would play a major role in his theory of the evolution of man. As a mature evolutionist thinking back over his own experiences, especially those in Tierra del Fuego, Darwin tried to rank men as belonging to different gradations in the varying stages of evolution—and hence took them to be manifesting different stages of developing humanity. Darwin set the difficulty of human classification in broad evolutionary context. His theory led him to rank races, and speak, with dry detachment, about human racial competition and extinction as evolution-in-action: as with ape to man, savage to Englishman, so with man on his way to something else struggle and annihilation was the only way up the evolutionary ladder. And on what rung were the Fuegians? He judged that "in this extreme part of South America, man exists in a lower state of improvement than in any other part of the world."[26] As he would later come to view it, people like the Fuegians or "the negro or Australian" were something like intermediate species, less evolved from the ape, and hence more likely to lose in the relentless struggle of the fit against the unfit. Darwin seemed only dimly aware that his theory and his sincere humanitarian compassion, which expressed itself in his hatred for slavery, might be in some sort of essential conflict. It was an ambiguity he carried within him throughout his life, and one he bequeathed to us as a legacy.

The *Beagle* dropped Jemmy, York, Fuegia Basket, and an Anglican missionary named Richard Matthews among Jemmy's tribe (the Yahgan), a little west of where the *Beagle* had collected the Fuegians on its previous voyage. (York and Fuegia were of the Alakaluf tribe, closer to the *Beagle*'s original landfall.)

The missionary effort was a failure, at least in the short term. When the *Beagle* came back a few weeks later, Matthews came screaming in terror and relief to the ship. Jemmy, York, and Fuegia had already begun to slip from their temporary dip into Anglicanism back into the ways of their tribes, but they promised to try to carry on the missionary effort without Matthews.

The *Beagle* set sail for the Falkland Islands in February 1833. After that, she would backtrack up the East Coast of South America so that Captain FitzRoy could fine-tune his survey. The plan was to revisit the mission at Tierra del Fuego the following summer before heading west into the Pacific.

At the Falklands, the *Beagle*'s crew was surprised to see the Union Jack flapping in the breeze. British warships, unbeknownst to FitzRoy and his men, had seized the islands from Argentina in January. The *Beagle* was welcomed as naval reinforcement until more British ships could arrive. FitzRoy meanwhile bought a schooner with his own money—a seal-hunting ship, (the *Unicorn*, which he renamed the *Adventure*), but which needed refitting with a new copper bottom to protect it from wood-boring worms—to accompany the *Beagle* on its mission. It would not be used for seal-hunting, but to give FitzRoy another vessel to help chart the coast of South America. From the Falklands they sailed to Maldonado, Uruguay, where the *Adventure* could receive its repairs. Darwin, during his time on shore, happily mixed it up with the local Spanish landowners and Gauchos. His guides were "well-armed with pistols and sabres," which Darwin naively thought unnecessary, until he learned of a recent traveler who ended up by the side of the road with his throat slit.[27]

Darwin was as amazing to the locals, as they, in their relative rudeness of civilization, were to him. His technology was, to them, magic—telling directions with a compass and striking matches with his teeth. A woman, sick in bed, and hungry for wonder, begged that Darwin would come and show her this strange magical instrument that could allow him, "a perfect stranger," to be able to point out the directions "to places where I had never been." And the miracle that "a man should strike fire with his teeth" was so astounding that whole families would gather to watch Darwin light a match—he was even offered "a dollar for a single one." If that weren't enough to astound the natives, they were entirely thrown out of sorts by Darwin's habit of washing his face every morning.[28]

The Gauchos were as magnificent as they were brutal, "tall and handsome," amply mustachioed and long-haired, with a "proud and dissolute countenance." They were, Darwin noted, exceedingly polite and graceful, but at the same time, "they seem quite as ready, if occasion offered, to cut your throat."[29] Up for any adventure, he tried his hand with the Gauchos' *bolas*—two balls tied together with a strip of leather—that the South American cowboys used to entangle their prey. "One day as I was amusing myself by galloping and whirling the balls round my head," one of the balls "struck a bush; and its revolving motion being thus destroyed, it immediately fell to the ground, and like magic caught one hind leg of my horse. . . . The

Gauchos roared with laughter; they cried they had seen every sort of animal caught, but had never before seen a man caught by himself."[30] Happily, he could always shoot dinner rather than snare it with *bolas*, and his training in the Glutton Club at Cambridge prepared him well for dining on ostrich, armadillo, and other South American fare.

The *Adventure* was ready to sail by mid-July, but Darwin decided his time was better spent on land. So he took off on horseback, promising to meet the *Beagle* on its return down the coastline. His days were spent hunting, skinning, and preparing specimens to pack off to Henslow, and nights, singing, smoking, and sleeping under the stars with the Gauchos "as comfortably with the Heavens for a Canopy as in a feather bed."[31]

The expedition was not all peace and sublimity, though. As Darwin quickly learned, he was exploring the interior in the midst of a war between the indigenous South American Indians and the conquering Spaniards. The Spanish were brutal in their conquest, but Darwin judged the Indians to be even more brutal in their defense and retaliation. That wasn't the sole source of political unrest. Having wound his way back to Buenos Aires, he found the city in turmoil, the Spanish military at odds with a warlord's rebels. He made his escape, and met up with FitzRoy again at Montevideo, where he was able to spend more peaceful days packing off specimens for Henslow—about two hundred animals skins, some mice, fish, insects, stones, seeds, and fossils.

Since self-preservation and science were so intimately linked during Darwin's daring exploits inland, he sometimes confused the two, eating a particularly important specimen, a new species of ostrich, rather than pickling it and crating it up for Henslow in England. Happily, the state of anatomical science was sophisticated enough back home that the little left over from the meal was enough to identify it, and then aptly name it *Rhea darwinii*.

Darwin was after both the living and the long dead, and especially those fossils that linked the two. One of his great finds was a partial *Megatherium* skull, an extinct version of the living land sloth, only much larger, and a llama-esque or camel-esque, long-snouted *Macrauchenia patachonica*.

Charles wasn't the first fossil hunter in his family, interestingly enough. For that, we must go to his great-grandfather, Robert Darwin, the corpulent barrister of Elston Hall. While Robert was no naturalist or scientist, he did discover a 200,000,000-year-old fossil of *Plesiosaurus dolichodeirus* by the well at the Elston Rectory.[32] Robert offered it to the Royal Society on December 11, 1718. Today it is displayed at London's Natural History Museum. This specimen of the prehistoric marine reptile was only about ten feet long, although the Plesiosaurus could be over fifteen feet. In his paper published on the find, William Stukeley of the Society surmised that it was "a rarity, the like whereof has not been observ'd before in this Island," and guessed that "it cannot be reckon'd Human, but seems to be a *Crocodile* or *Porpoise*."[33] Stukeley was not too far off, given the partial nature of the fossil skeleton. In reality, *Plesiosaurus dolichodeirus* looks more like a bloated giraffe with fins.

In March, the *Beagle* revisited the mission in Tierra del Fuego, a miserable experience. Jemmy was almost unrecognizable, and ashamed of his appearance. He pulled up beside the *Beagle* in a canoe and told how Fuegia Basket and York Minster had stolen all his possessions, including his English clothes. All that remained to him was his wife. That, however, was enough; he was contented to stay with her, living as a savage, stripped of the veneer of Anglican civilization.

Leaving Jemmy behind, the *Beagle* sailed back to the Falklands, only to find that rebellious gauchos and Indians had joined hands in murdering the Englishmen charged with holding the islands. The rebel leader, one Antonio Rivero, was captured by the *Beagle*'s men, and clapped in chains. While on the islands, Darwin received mail, informing him that his specimens, carefully placed by Henslow, were now the talk of English scientific society—it was Darwin's first glimpse of his future fame as a naturalist.

Leaving the Falklands, the *Beagle* traveled west to Rio Santa Cruz on the Argentine mainland, and then down and around the cape and into the Pacific Ocean in early June.

Darwin was as captivated as ever by his surroundings, viewing every geological formation, every bit of flora and fauna, through Lyell's *Principles of Geology*, and surmising that all such objects or living things were the result of constant, imperceptible, natural changes. These observations became, in retrospect, acute when the *Beagle* reached the Galapagos Islands in September 1835.

At the time, Darwin's visit to this small set of volcanic islands did not seem so very momentous—but he was certainly captivated by his discoveries there. On these relatively bare islands, hundreds of miles from the South American mainland, Darwin found unique variations of creatures that could also be found on the continent. Even stranger, these variations varied themselves according to each island. The famed Galapagos tortoises have such distinct shells that those familiar with them could tell which of the islands they came from by their shape, and the even more famous finches could be distinguished, island for island, by the shape of their beaks, each aptly fitted to the food it ate.

Actually, Darwin completely missed these distinctions and their significance while on the Islands. There was no grand evolutionary aha! at Galapagos. His own grandfather had provided him with a perfectly interesting reason for the differences, and hence a reason to look for them carefully. For Erasmus, each kind of finch (for example) would strive to feed itself under its own peculiar conditions, thereby bringing about slight transmutations over time. But here on the Galapagos, something didn't quite fit Erasmus's theory. The islands presented what amounted to nearly *identical* environments. It would make sense that finches on the bare volcanic islands of Galapagos would differ from those found on the west coast of South America: different environments, different transmutations. But why should there be differences on nearly identical islands a mere twenty miles apart? It was in fact because of their very proximity, as Darwin humbly admitted, that he didn't keep very straight what specimens came from which island.[34]

And how did they get there? Were they just plopped down from heaven? Did some great geological catastrophe transport them to separate islands, flinging them up on the various Galapagos beaches like so many Robinson Crusoes forced to make do under very different conditions than they had experienced on the continent? Maybe it was the reverse, and the islands had somehow been connected to South America, and the land in between was slowly worn away? Or was it quickly destroyed?

Darwin was probably too busy investigating and collecting to do much deeper reflection on these points. The survey of the Galapagos Islands was slated for only a few short weeks until they set sail for Tahiti in the third week of October. He would have to take what he could, and think about it later.

As compared with the Fuegians, the Tahitians were of far better stock. "I was pleased with nothing so much as with the inhabitants. There is a mildness in the expression of their countenances, which at once banishes the idea of a savage; and an intelligence, which shows they are advancing in civilization."[35] No doubt it helped that the natives had already been evangelized by Christian missionaries, knew a little English, and didn't greet the *Beagle* with menacing, barbaric ululations. Darwin thought their higher state of civilization should be credited directly to Christianity.[36]

From Tahiti, it was off to New Zealand (which he disliked). It was the end of November 1835, and by now Darwin just wanted to go home. The aboriginal New Zealanders, the Maori, were also a disappointment; Tahitians they were not. While one takes them to be "belonging to the same family of mankind," yet a glance at the New Zealand Maori reveals that he is a "savage" while the Tahitian is "a civilized man."[37] The Maori has "a twinkling in the eye, which cannot indicate any thing but cunning and ferocity," and "their persons and houses are filthily dirty and offensive: the idea of washing either their bodies or their clothes never seems to enter their heads."[38] Darwin had cheerfully endured a good mutual chest-thumping as the greeting of native Fuegians. He found the conventional greeting of New Zealanders much more disconcerting. These particular natives went in for pressing noses hello while "they uttered comfortable little grunts, very much in the same manner as two pigs do, when rubbing against each other."[39] Forever amiable, Darwin joined in, recalling with a sigh the Reverend Henslow's advice about being congenial among the unwashed.

Charles spent his fifth Christmas of the voyage in New Zealand, "and the next, I trust in providence, will be in England." Darwin attended church where "part of the service was read in English, and part in the New Zealand language." Whatever was redeemable about the Maoris, Darwin again put down to Christian missionary efforts.[40] When Darwin made these observations it was not because he was growing in Christian devotion, but because as an heir to Enlightenment Deism, he could easily believe and affirm that the chief good of religion was the improvement of the barbarous and ignorant; while Deism tried to strip Christianity from its doctrinal truths, it did not deny its moral content or disallow that, from an entirely secular point of view, it could produce good effects. Darwin, in his observations of the more Christianized and the lesser Christianized tribes, believed he saw these effects—indeed, he thought the differences were stark and indisputable.

From New Zealand the *Beagle* tacked to Australia (which he liked, because it was much more like England than any other destination of the trip). He found the "black aborigines" to be not as savage as he'd heard. They were surrounded in Australia by a larger, white civilization that Darwin believed was on the way to extinguishing them, in part because of the introduction of European diseases.[41] He also reckoned that the kangaroo and the emu were doomed to extinction as well, and he did his part by going kangaroo-hunting. "It may not be long before these animals are altogether exterminated, but their doom is fixed."[42]

From Australia, they hurried through the Indian Ocean, then around the Cape of Good Hope, and made a short stay along the eastern coast of South America before sailing back to England, arriving on Falmouth on October 2, 1836, completing his journey just shy of five years. It was almost a shock to be home after so much time away; but the greater shock was that through the efforts of Henslow and Sedgwick, who had trumpeted his finds sent from the *Beagle*, Charles Darwin was already a very famous scientific man.

Chapter 4

Hatching the Evolutionary Plot

When Charles surprised his family in the breakfast room at the Mount after returning from his voyage, the first words out of his father's mouth were, "Why, the shape of his head is quite altered." On the face of it, a rather odd thing to say. The truth of the matter, one suspects, is that the shape of his head was now rather more visible, given that Darwin was well on the way to baldness. But Darwin in his *Autobiography* puts the remark in a more interesting context—namely, that of phrenology.

We might think that phrenology, reading the bumps on the head as indications of a person's character and abilities, is a pre-scientific pseudo-science that superstitious folk in the nineteenth century clung to in the same way that some people today hang onto the astrology page in their local newspaper. Such is not the case. Phrenology was the most advanced science of the materialists of the time. Since the mind was merely material, the contours of the head betrayed the physical foundations of one's entire psyche. Phrenologists, then, were not the astrologers but rather the neurologists of their day; or at least they were assumed to be the nineteenth century's leading specialists on the brain.

While Robert Darwin was "far from being a believer in phrenology," even he was at least momentarily convinced that Charles's head, made visible by the subsidence of his hairline, revealed a great intellectual metamorphosis. In his *Autobiography*, Charles Darwin says that the changing shape of his head gave evidence of his advance from a brute to a scientist. Indeed, Darwin believed the changes in his head had come through his long hours thinking aboard the *Beagle* in the same way that (as his grandfather and Lamarck would have it) over generations the giraffe's neck grew longer from stretching to eat leaves high up in the trees. During the first two years of the voyage, "my old passion for shooting survived in nearly full force," but he soon gave up the gun to his servant, "as shooting interfered with my work, more especially with making out the geological structure of the country." One pleasure yielded to another so that soon "the pleasure of observing and reasoning was a much higher one than that of skill and sport. The primeval instincts of the barbarian slowly yielded to the acquired tastes of the civilized man." Darwin, the young gentleman delightedly picking off partridges, and Darwin the semi-wild Gaucho riding madly over the scrub plains hunting down the evening supper, were now extinct, replaced by Darwin the scientist, with his prominent head pushing up through the receding hairline to prove it.

Darwin therefore saw himself as recapitulating in five years the evolution of human beings from savagery to civility, and with visible, material results. "That my mind became developed through my pursuits during the voyage, is rendered probable by a remark made by my father, who was the most acute observer whom I ever saw," that remark again being about the shape of his head.[1]

Darwin wrote to Captain FitzRoy about his homecoming: "My sisters assure me I do not look the least different"—not apparently noticing his transformed head—yet, he added, "all England appears changed, excepting the good old Town of Shrewsbury & its inhabitants—which for all I can see to the contrary may go on as they now are to Doomsday."

As G. K. Chesterton later illustrated so marvelously in his novel *Manalive*, the best way to see again the true splendor of our own home is to travel around the world and come back to it. Darwin confided in a letter to his former Captain, "As I drew nearer to Shrewsbury everything looked more beautiful & cheerful—In passing Gloucestershire & Worcestershire I wished much for you to admire the fields, woods & orchards—The stupid people on the coach did not seem to think the fields one bit greener than usual but I am sure, we should have thoroughly agreed, that the wide world does not contain so happy a prospect as the rich cultivated land of England." He signed off, playfully, "Good bye—God bless you—I hope you are as happy, but much wiser than your most sincere but unworthy Philos. Chas. Darwin."[2]

FitzRoy, who had done so much for both Darwin and England, would feel himself more and more in the shadow of

Darwin's rising fame over the coming years. This was coupled with a feeling of remorse that he had aided and abetted the formation of Darwin's evolutionary theory, and it seems FitzRoy's future turn to strict biblical literalism was in no small part a penitential reaction. On top of all this, FitzRoy felt ill-appreciated for his services to the crown. Like Darwin's own Uncle Erasmus, he would later crumple into suicide under the combined pressures he felt weighing upon him.

But no such future darkness ruined Darwin's triumphant return. He was indeed a changed man. There would be no parsonage in his future, nor would he try to combine religion and science as a university priest-professor like Henslow or Sedgwick. He was a naturalist, not a supernaturalist, and he came off the ship bent like a bow to do everything in his power to make his mark in the world of science.

As could be expected, Charles was immediately captured, tugged, and pulled this way and that by family and friends, ricocheting from one house to another in a welcoming frenzy that had Darwin saying in happy exaggeration that the "busiest time of the whole voyage has been tranquility itself to this last month."[3] It was a drastic change, exhausting and satisfying, a feast of family for a man who had been starved for home.

Despite the pull of his family, Darwin was determined to sort out everything he'd gathered during his five-year adventure: boxes of exotic specimens, a ship's journal that had to be made into a book, geological observations and speculations that had to be set before the great men of British science.

He lost no time in trying to get things aligned, writing to Henslow the day after arriving at the Mount, and admitting that he sorely needed "advice on many points, indeed I am in the clouds & neither know what to do, or where to go." He could not do it alone, as he clearly realized. He would need a small army of experts willing to spend time on *his* project, friends in high places willing to push *his* cause. Darwin was quite willing to use his characteristic, natural, humble charm to ensure that such help was forthcoming. "My dear Henslow, I do long to see you; you have been the kindest friend to me, that ever Man possessed.—I can write no more for I am giddy with joy & confusion."[4]

Once again, the Reverend Henslow had been more than a friend to Darwin. Not only had he done the initial sorting of the specimens sent by Darwin, but he had carefully built up Darwin's reputation as a new and eminent scientist. To that end, he'd circulated some of Darwin's unpolished writings, thoughtfully but hastily scribbled while on the *Beagle*.

Darwin was delighted to discover that, thanks to Henslow's efforts, the great geologist Charles Lyell—for whom Darwin now had an almost worshipful admiration—was eager to count him as a disciple. Lyell saw Darwin as an ally against men like Sedgwick, and Charles was deeply gratified when Lyell graciously admitted that Darwin's account of the development of coral reefs was superior to his own. Darwin and Lyell became close friends. Like Henslow and Sedgwick, Lyell was "very kindhearted," but unlike them, he was, in Darwin's own description, "thoroughly liberal in his religious beliefs or rather disbeliefs; but he was a strong theist,"[5] in other words, he was a Deist Whig, a type with which Charles was well-familiar and happy.

Darwin soon landed another great catch with his splendid supply of fossils: the stout Tory Richard Owen. Owen was the most important and brilliant taxonomist in Britain, rivaling an earlier genius, the great Frenchman Georges Cuvier. Both men were masterful anatomists. They could take an isolated bone, and from its type and shape, divine the creature to which it must have belonged. They could perform this intellectual magic because they understood that all parts of each distinct creature were intimately formed according to the whole.

Cuvier had done more than anyone to quash the popularity of evolutionary theory as outlined by Lamarck and Erasmus Darwin, and on these very grounds: creatures were formed wholes; parts of their bodies could not change without changing the entire animal; bones and organs were not interchangeable, which was why, with his expert knowledge of anatomy, Cuvier could discern the entire structure and habits of an animal from a single bone. Cuvier and Owen were well aware of the fossil evidence displaying layers of similar creatures caught in layers of geological time. Indeed, Cuvier's extraordinary anatomical powers were what had largely determined that such measured extinctions had occurred. But because the arrivals were just as sudden in the fossil record as the extinctions, Cuvier could see no evidence of slow and smooth transformations of one creature into another as the transmutationists were demanding.

Cuvier had been the first to indentify the extinct creature *Megatherium*, and now Richard Owen was quite happy to receive the *Megatherium* remains discovered by Charles Darwin. There were not many such remains at hand. Owen helped Darwin identify bones and the creatures they came from; and some of the fossil bones in Darwin's collection were previously unidentified. From bits and pieces of skeletons provided by Darwin, Owen reconstructed an enormous, extinct, sloth-like creature, *Scelidotherium*, weighing in at about three tons, and the *Toxodon* and *Glyptodon*, oversized versions of the capybara and armadillo, respectively.

But Darwin needed more than bones sorted out. He was trying to piece together how everything he had seen, collected, and thought fit into a grand evolutionary vision; he was full of ardor and ambition to make his mark. He knew he had found his niche. As his predecessors had found their fame in medicine or law, he would make his as a naturalist (and in so doing would vindicate himself in the eyes of his father, or so we can reasonably speculate). Darwin's dedication to this work demanded that he tightly control his time. He could not be the same lovable homebody his family had known before, ever up for a good meal and an evening of comfortable conversations around the fire. He soon rented a house in Cambridge, in part to keep his family at arm's length, and then moved to London. Here, he could be near his dear brother Erasmus, now an effete bachelor and socialite among the radical set. Rather than practicing medicine, Erasmus lived off his father's money. He was the most open and lovable skeptic of the Darwin family, a sparkling wit, with endless effusing charm. He was a copy of his grandfather Erasmus, only much thinner and entirely devoid of ambition.

Charles, on the other hand, burned with ambition, and with so much to do, he wanted to do everything at once. The most important task, undertaken in the first nine months of 1837, was turning his *Beagle* journal into a naturalist's travel adventure. FitzRoy had suggested this to Darwin during the journey, and had Darwin never written anything else, his *Journal of*

Researches into the Geology and Natural History of the Various Countries Visited by H.M.S. Beagle (as it eventually came to be titled) would count as a classic of its genre; even more interesting, if this had been his only work, we would never connect Darwin with evolution, because there is not even a hint of "transmutationism" in its pages. Had Darwin come out as an evolutionist, he might have upset his friendship with Lyell—and Lyell was leading him upwards among England's scientific elite. He was also leading him in other interesting directions. While Darwin was at sea, Lyell had married one Mary Horner. Now with an apparent queue of Horner daughters waiting to be married, Lyells and Horners shuffled, whispered, plotted, and conspired to bring this most eligible and busy bachelor to marriage. They failed. Instead, Darwin reinforced the Darwin-Wedgwood alliance. His father had married Susanna Wedgwood, daughter of Josiah Wedgwood I. Darwin married his own first cousin Emma Wedgwood, the daughter of Uncle Jos (Josiah Wedgwood II) at the beginning of 1839.

The courtship was as unromantic as one could imagine. No strutting in front of Emma showing his plumage, no fond and furtive looks from across the great dinner table at the Wedgwood estate, no singing under the window like a lovesick bird. It was all too businesslike. Quite simply, Darwin needed a wife.

Darwin had actually taken a sheet of paper and tallied up the pros and cons of marrying. Given all the anxiety he had over his work, marriage barely eked out a victory. Emma was no exotic beauty, and Darwin was an admittedly plain-looker. They were very well known to each other, as the tight Darwin and Wedgwood connection stretched back two generations. There can be little doubt that the familiarity was a strong part of Darwin's motivation in proposing.

The proposal turned out to be little more than a disaster. Already experiencing the characteristic sickness that would be his constant companion, especially when faced with some anxious situation, he trudged to the Wedgwoods' home and asked for Emma's hand. Darwin seemed to go about it as a hard but necessary task. For Emma it was a surprise, but not a romantic one. She was taken aback precisely because she had known him all her life, and there had never been the slightest twinkle of interest in her cousin's eye. The wedding was lackluster as well. Darwin had a throbbing headache. Their honeymoon consisted of the carriage ride to their new London home, sandwiches and water included.

Despite this grey and discomfiting beginning, Charles and Emma would grow to love each other deeply, and Charles turned out to be the gentlest of husbands. Contrary to his cool calculations about how much of a bother it would be to have children interfering with his work, and even more, his brooding about the expense, he was an affectionate, doting father, an incurable baby kisser, a regular fountain of saccharin endearments. Unlike his stern father, he melted to butter and sugar in front of his children, especially when they were sick. His humility and kindness to his servants (and even animals) were singular and exemplary. We may have a notion of him as a scientist who was all head and no heart, but we have not peered into the windows of his home to see him playing backgammon for hours to help a sick daughter pass the time, or listening dutifully to Emma play the piano every night, even though his ear was entirely of tin. In short, he was a model family man and gentle-man.

He was all these good things, and yet, he was entirely devoted to his work, his theory, and he informed Emma from the very beginning that the Darwin household, as it grew, would be bent and subservient to one immense, all-embracing task: the vindi-cation of his theory of evolution. He longed and lived to bring to complete fruition the seed that his grandfather Erasmus had planted.

Although that might not be quite the way we should put it. One of Charles Darwin's very few character flaws was this: he was oddly possessive about *his* theory, so much so that he failed to acknowledge his predecessors, including his own grandfather, until his detractors pointed out the glaring omissions. He wanted the theory of evolution to be *his* discovery, *his* creation, *his* baby. He was, to say the least, single-minded in the intensity of his devotion.

This devotion had no religious aspect. He was his father's son. Against his father's explicit warning not to let the women folk know his true beliefs (or unbeliefs), Charles had been admirably frank with Emma. She was distressed, and with some hesitation asked him if he would read, for *her*, "the end of the 13th chap of St John," the part of the New Testament she loved best. In the King James version it reads,

> Therefore, when he was gone out, Jesus said, "Now is the Son of man glorified, and God is glorified in him. If God be glorified in him, God shall also glorify him in himself, and shall straightway glorify him. Little children, yet a little while

I am with you. Ye shall seek me: and as I said unto the Jews, Whither I go, ye cannot come; so now I say to you. A new commandment I give unto you, That ye love one another; as I have loved you, that ye also love one another. By this shall all men know that ye are my disciples, if ye have love one to another."

This was not a vague admonition to kindness for Emma, something that even a Unitarian could agree with while politely letting go of the divinity of Christ. Emma was a different creature than her skeptical father—and ironically she had her father to blame for it. Like Robert Darwin, Josiah Wedgwood had fallen far below the featherbed of Unitarianism. But, in a bow to maintaining social order, he respectfully sent his children to attend Anglican services. In doing so, Emma ended up swimming upstream against the two-generation current of Enlightenment skepticism and reaching the old shoreline of Christianity, or at least an eroded part of it, which gave her a fairly latitudinarian grasp of Christian doctrine. Nevertheless, she had, in her way, made a real journey to faith. Darwin, on the other hand, had let the current of Enlightenment disbelief— so prevalent in his family and its social circle—carry him along. Emma worried about the fate of Darwin's soul. He was convinced he didn't have one. Her world was so large that it extended into the next. His world was contracting daily, so that even the most cherished things they both took for granted would soon be squeezed out.

Not too long after they were married, Emma wrote Charles another letter, a rather odd thing to do we might think, since they lived in the same house, but it allowed her to express her worries clearly and give him tangible words over which to brood. In it, she hit dead-on the problem of his single-minded intensity. "You are so absorbed in your own thoughts," Emma warned him gently, that it is "very difficult for you to avoid casting out as interruptions other sorts of thoughts which have no relation to what you are pursuing or to be able to give your whole attention to both sides of the question."

For Emma, the "side" that received short-shrift was God. Darwin had no healthy fear of disbelief (or the wrath of God), and this, she thought, was unnatural. She hinted that it was his charming, urbane, and beloved brother Erasmus, whom he so admired, who killed this fear through his sarcastic wit. While even his father had encouraged Charles to become an Anglican divine, his brother Erasmus had doled out cheerful scorn. Emma recognized that Erasmus's good opinion meant a great deal to Charles.

"It seems to me," continued Emma, "that the line of your pursuits may have led you to view chiefly the difficulties on one side, & that you have not had time to consider & study the chain of difficulties on the other, but I believe you do not [yet] consider your opinions as formed." Were they fully formed even now? How much had Darwin actually revealed?

We don't know. But we do know that Emma longed for the two to be of one mind and heart as a married couple, and even more, not to be separated by different eternal destinations. Darwin was genuinely touched in his heart but not his head. "When I am dead," he wrote on the outside of the letter, "know that many times, I have kissed & cryed over this."[6] Yet for Darwin, the notion of a soul and the afterlife was by now entirely unintelligible. He was a thoroughgoing materialist, just as his grandfather had been, just as his father remained.

We know this because for about two years he had been busy writing away in his very private notebooks, all his most private thoughts about transmutationism. And the notebooks make very clear that he was after a particular version of the transformation of species, an entirely materialist version, one that began, with the aid of his father, as a meditation on his grandfather's *Zoönomia*. In his "M Notebook" of 1838, we find that he probes his father for information, and both are bantering back and forth about the *Zoönomia*. Again and again we find "my father thinks," "my father says."

Evolution was a family affair, yet it was to be *his* theory, profoundly materialistic and curiously designed not to let a divine foot in the door. From the very opening pages of his "Notebook B" on transmutationism, which he began in the summer of 1837, we find him aiming directly at an entirely reductionist explanation of human beings; every aspect of them was to be accounted for entirely as the result of their animal origins, even their ability to think. As he wrote somewhat cryptically, "Each species changes. does it progress. Man gains ideas."[7]

What of moral character? There is no need to raise man to a different moral, and hence metaphysical plane than the animals. Even here, he is no different. The passions of human beings and animals are on one continuum, and these passions are just what morality is about. "Origin of man now proved.—Metaphysics must flourish.—He who understands baboon would do more towards metaphysics than Locke A dog *whines*, & so does man.—dogs laughs for joy, so does dog bark. (not shout) when opening his mouth in romps, smiles."[8]

Free will? In the "Notebook M" he wondered whether jolly or ill-tempered fat men are jolly or ill-tempered because they are fat? "Thinking over these things, one doubts existence of free will every action determined by heredetary [sic] constitution ... I verily believe free-will & chance are synonymous."

And so we come back round to thought, and even more, discovery. Do we freely inquire into nature? Are our insights the result of our efforts? "Shake ten thousand grains of sand together & one will be uppermost:—so in thoughts, one will rise according to law."[9] One wonders about the status of this particular thought. Could the laws of evolution, shaking the minds of men, bring the right theory to the surface at the right time through the labors of the right man?

What of religion? Did Emma know what Charles thought about *her* faith? "It is an argument for materialism, that cold water brings on suddenly in head, a frame of mind, analogous to those feelings, which may be considered as truly spiritual."[10] Poor Emma. So much for her spiritual intimations and worries, they were of a piece with cold showers, apparently.

Darwin meditated not only on life, but also death, for everything must be explained as caused by evolution, even the limited life span of each creature. On one entry, perhaps unconsciously echoing Descartes' famous dictum, "I think, therefore I am," Darwin merely scrawled "I think," and then under it, put a small branching bush, a single thin trunk splaying off abruptly into three, two of which blossom with new varieties bursting forth from each new node. *I think, therefore, it is*—the grand vision of evolution unfolds. He later added, "REQUIRES extinction,"[11] and further on, "Heaven knows whether this agrees with Nature: *Cuidado*." "Care" in Spanish.[12]

Indeed, death became the strange focus of his theory, the dark creative force he embraced to make God unnecessary. In September 1838 Darwin picked up that great macabre masterpiece written at the end of the eighteenth century, Thomas Malthus's *Essay on the Principle of Population*. Malthus argued that an ugly "fact" of life was that the more you feed people, the more they will breed, and so there will always be far more mouths to feed than food to put in them. The poor will be with us always, and any attempt to help the miserable wretches will only produce more misery by creating even more wretches. Better to let the excess population die, mind your own business, and work hard. The poor, of course, should be worked harder so that they might find virtue in the workhouse and also have less time to breed. Malthusianism was cheerfully accepted by many secularist, liberal members of Britain's industrial, professional elite—the very class of the Darwin-Wedgwood alliance. Malthus's theory seemed scientific, practical, and moral all at the same time—or at least it gave a moral covering to the interests of the industrialist and the eugenicist; and eugenics was the natural direction of the non-theist evolutionist.

Not all believed the Malthusian charter—a most striking example being Charles Dickens. His Ebenezer Scrooge is Malthusian man made flesh. "If the poor are going to die without my charity," says Scrooge, before his great transformation, "they had better do it and decrease the surplus population." Later, when a repenting Scrooge tearfully begs to know if the sickly cripple Tiny Tim will live, the ghost of Christmas Present throws Scrooge's cold words back at him: "If he be like to die, he had better do it, and decrease the population." The ghost then indicts Malthusianism:

> Man, if man you be in heart, not adamant, forbear that wicked cant until you have discovered what the surplus is, and where it is. Will you decide what men shall live, what men shall die? It may be that in the sight of Heaven you are more worthless and less fit to live than millions like this poor man's child. O God! to hear the insect on the leaf pronouncing on the too much life among his hungry brothers in the dust!

Darwin, however, did not have the benefit of being visited by the ghosts of Christmas; and for him, Malthus was far too useful to be dispensed with. Malthus allowed Darwin to put death in a new, more positive light. Life was profligate, imprudently overproducing, casting forth far more than it could ever feed, generating every manner of variation of species, with no thought about how to care for them all. But death was a good accountant; it knew how to deal with "too much life." It cut down the surplus population, and did so with ruthless precision. The weak, sickly, malformed, unfit, all were methodically removed by its cold, keen scythe.

Death, Darwin thought, was the key to life, a complete inversion of Emma's superstitious belief in a creator God and the idea that death was the punishment for original sin. Death was, is, and always will be, the creator. Unlike the biblical God, it does not pronounce everything good, it does not demand peace; instead it is the winnower of dross and imperfection, and by this means of culling surplus populations it creates a fitter species. War, the incessant struggle of creature against creature, species against species, is the true furnace of creation and progress.

All this had been fairly well worked out before Charles and Emma moved into their modest home in London at the end of January 1839, united as husband and wife, one flesh but holding two decidedly different visions of the cosmos. Darwin threw himself into setting up their house with the giggling zeal of a newlywed, "like an overgrown child with a new toy."[13] Darwin was, as always, a collector, and set about in earnest on the hunt for pots, pans, dinnerware, furniture, maids, butlers, all the endless necessities and accouterments of a gentlemanly house, albeit a modest one. Presents arrived, and then there were Darwin's books and papers. The house filled up quickly.

Even though Emma had known Charles her entire life, living with a man as husband is different from knowing him as a familiar cousin. His obsession with work was a trial, and Emma had to live with her husband's strange, incessant headaches and nausea. The obsession seems to have been, in great degree, the cause of his malady, the pressures of work bringing about anxiety and fatigue, and triggering the bouts. The upside of his bouts of reeling and vomiting was that Charles knocked off work, and allowed himself—indeed, happily indulged in—Emma's affectionate nursing. If it weren't for this, they would never have been so close. Otherwise, it was up at seven, writing until breakfast at midmorning, then more work—with Emma sometimes quietly doing needlework in Charles's study—then a good walk, some attention to necessary business, dinner at six, and finally, a dutiful listening to Emma on the piano in the evening. Occasional dinner parties punctuated the routine, as well as visits by Darwin to his scientific societies and friends.

When Emma got pregnant, it was her turn to be nauseated and call for her husband's ministrations, which were sincerely given when Darwin himself wasn't retching. Poor Emma, Darwin's problems seemed to flare up as kind of a sympathetic response to her morning sickness. William Erasmus Darwin was born in December 1839, and no father could have been more proud, more embarrassingly affectionate, and more convinced of his son's endless excellences and hidden talents. A daguerreotype survives from 1842 of Charles and young William, a smugly beaming father who knows, to his complete satisfaction, that he holds a once and future king on his lap. "Doddy," he called his son; but certainly not many fathers would search their child's face, as Darwin did, for signs of his evolutionary animal origins.

Darwin was, by this time, a famous man, the author of the *Journal*, the great *Beagle* adventure, who was set to climb much higher on the scientific ladder of success. But in the daguerreotype, Darwin's eyes reveal the toll. He was sicklier, his eyes were dark and haggard, and his weight had dropped below one hundred and fifty pounds. Not good, for a man almost six feet tall. He seemed to be singularly unfit to survive the rigors of science and the pressures of secretly fine-tuning his arguments about the survival of the fittest.

The next baby to come was Anne Elizabeth, in March 1841. No doubt Darwin was momentarily perplexed that having a girl, he would have to sort through other possible names than

Erasmus, Robert, or Charles. Inevitably, the anxiety of it all, on top of his work, struck him down, and so he went off to recuperate at the Wedgwood estate, and then on to the Mount for his father's medical advice. The ironic clash of his life and his theory was not entirely lost on him. "It has been a bitter mortification for me to digest the conclusion that 'the race is for the strong'—and that I shall probably do little more, but must be content to admire the strides others make in science."[14]

His health returned in 1842, allowing him to write out in more detail his theory on the transmutation of species, and here we find all the basic arguments present in Erasmus Darwin's *Zoönomia*, but with the addition of a death-dealing natural selection, which replaced Lamarck's assumption about an organism's inner strivings leading to physical transformations. Darwin also took account of the objections by Lyell and others—objections that species modifications were strictly limited, that the fossil record did not record smooth transitions between species, and that there were no intermediate forms existing as living proof that such transformations were occurring. These remained as just some of the obstacles against which his own theory would have to struggle to survive. As with his later *Origin*, he met them, not by arguing against them directly, but largely by appeal to indirect evidence or circumstance that would explain away the problem. The fossil record, for example, was not yet fully unearthed, and further, fossils simply were not that frequently preserved, so gaps were only gaps in fossilization, not in actual species that had lived.

Darwin and Emma both longed to be back out in the country, a considerably healthier atmosphere than London could provide, no small concern since Emma was with child again. Backed by a significant inheritance, they went house-hunting in the summer of 1842, two sparrows looking for a good nesting site. They settled on a dowdy, thickset house at Down (or Downe, as it was later spelled), an eighteen acre property that would be Darwin's home for the rest of his life.

A little over a week after they moved in, Mary Eleanor Darwin was born. She would live three weeks, and then settle to her eternal rest in the graveyard at the church in Down. It is one thing to have an abstract theory about death, and mark out how it must divide the fit from the unfit, the strong from the weak. Death is then looking at others. But death had now turned its pitiless eyes upon Charles Darwin, and taken his third child.

To drive out grief, he threw himself into his work, and into becoming a proper English gentleman of his country estate. This wasn't just a matter of maintaining (or improving) the house and its landscaping, it meant he had local duties to perform—attending parish councils, taking an active role in the civic life of the village, and other social tasks that helped distract him from his mourning. But his chief consolation remained his work, the development of his theory, which he continued to keep private even while publishing papers on geology that would support but not reveal it.

His papers were, then, in some sense a tease. For instance, he confided rather vaguely and disingenuously to the botanist Joseph Hooker that "I am almost convinced (quite contrary to opinion I started with) that species are not (it is like confessing a murder) immutable," and that he had come up with a "simple way" different than Larmarck's "by which species become exquisitely adapted to various ends."[15]

Henrietta Emma Darwin was born at the end of September 1843, the first child to be born, and live, at Down House. She was nicknamed Etty, and would become a great help to her father, so much so that she edited out aspects of his *Autobiography* that she thought unsuited for publication, and removed worrisome passages from Charles's biography of his grandfather.

By the summer of 1844, Darwin's sketch of his theory had grown to a rather lengthy manuscript. Though only in his mid-thirties, Darwin feared death might suddenly take him before his work was finished. He wrote out a peculiar kind of will to his wife Emma, instructing her in morbid detail what to do with the manuscript if he should suddenly expire.

The sticky problem, however, was that all his eminent scientific friends—Henslow, Sedgwick, Lyell, and Owen—were intellectually antagonistic to transmutationism. All of them understood the fossil evidence very clearly, but had come to dramatically different conclusions from Darwin about it. They could hardly be faithful executors of his theory—especially because, as Darwin knew, he hadn't answered their objections to transmutationism. That was why, in great part, he kept working on his theory in private.

His theory! In November, Darwin experienced the quite comic shock of having a non-scientist, a popularizer, a mere journalist and publisher, come out with a book, *Vestiges of the Natural History of Creation*, which threw together the whole drama of transmutationism from the first swirls of matter that congealed to make the planets, and the first palpitating atoms that gathered themselves together spontaneously into a living mass, to the blossoming of each layer of now-fossilized plants and animals, and right up through the apes to man. Everything was there: astronomy, geology, botany, biology, paleontology. The *Vestiges of Creation* was an international sensation, going through edition after edition. Darwin had been scooped. How could this anonymous author have so thoroughly guessed at *Darwin*'s theory?

It is peculiar that Darwin should have been so surprised. His own grandfather's *Zoönomia* had spread the evolutionary news with a similar splash near the end of the eighteenth century. Lamarck's work had become even more widely known. Darwin's former Edinburgh colleague and mentor Robert Grant was a public proponent of the theory of transmutationism. Bits and pieces of Darwin's ideas had been expressed throughout the first half of the nineteenth century by men like James Prichard, William Wells, William Lawrence, Patrick Matthew, and Edward Blyth—and now by the anonymous author of *Vestiges* (who was in fact the journalist Robert Chambers, though his identity was not revealed until forty years later, in 1884).

In fact, the theory of evolution had been discussed much much earlier—at least as early as the first century B.C. That was when the Epicurean philosopher Lucretius provided the West with a full blown evolutionary account mapping out stage by stage how randomly justling atoms had produced everything in the universe up to man. Lucretius was the darling of the eighteenth century *philosophes*. He provided Deists and atheists with a thoroughly secular counterblast to the Christian scheme of creation and salvation. The intelligentsia had thus been soaked in evolutionary thought for nearly a century. It is, then, very odd, and quite amusing, that Charles Darwin should be shocked at the appearance of *Vestiges* and its popular success.

Darwin turned his shock into a kind of resentment. This popular work became a foil against which he had to prove his own originality. When he finally decided to go public with his theory, a decade and a half later, the shock was less about the ideas that were expressed than that so eminent a scientist as Mr. Charles Darwin, whom everyone thought quite reserved, could suddenly join ranks with such radicals as the evolutionists.

When *Vestiges* appeared, eminent scientists, among them Darwin's mentors and friends, felt called to refute its ideas. Sedgwick, for one, was incensed by *Vestiges*. He attacked it in the *Edinburgh Review*, nailing its unguarded assumptions and unwarranted leaps with such precision that Darwin felt the points sink into his own skin. In particular, Sedgwick argued that *Vestiges* was "intensely hypothetical," and the author, an obvious amateur, "builds his castles in the air, misconceiving the principles of science, or misunderstanding the facts with which it has to deal; or, what is worse still, distorting them to serve his purpose."

"I read it [Sedgwick's rebuttal] with fear & trembling," Darwin groaned.[16] The charge of spinning too little evidence into too grand a theory—that is, of betraying the proper model of inductive science—Darwin well realized, would be leveled at him. As he later wrote to Asa Gray in the year the *Origin* was published, "My work will be grievously hypothetical, and large parts by no means worthy of being called induction, my commonest error being probably induction from too few facts."[17] Beyond the confines of questions of science, Sedgwick was worried about the implications for morality and religion, commenting to Lyell in a letter that if the argument of the *Vestiges* were true, "the labours of sober induction are in vain; religion is a lie; human law is a mass of folly, and a base injustice; morality is moonshine; our labours for the black people of Africa were works of madmen; and man and woman are only better beasts!"[18] Whether Sedgwick had cause to worry, we shall see soon enough.

This much is certain. Darwin felt entirely deflated by the *Vestiges*. His grand theory had been scooped and then publicly skewered by prominent scientists—even friends. He retreated from sketching grand outlines to focus on the kind of research he had done under Grant. He threw himself into a small, neglected corner of nature to search for hints that his grand scheme was not mere speculative wind. He chose barnacles.

Darwin was an extraordinarily careful researcher, closely observing the bland and seemingly inconsequential Lilliputian biological terrain under the microscope until he got hints of a puzzle to be solved. He did nothing by halves, but halves of halves of halves until he thought he detected some monumental shift in the most minute detail. This habit, nurtured in his apprenticeship with Grant, was supported by his Lyellian bent in geology, where the smallest changes under our noses added up to moved mountains over time—if time indeed could be counted on to be in endless supply.

If, as Lyell's theory demanded, time's arrow pointed backwards indefinitely, then there was no beginning point that could interfere with slow changes adding up, age upon age, to anything imaginable. If a breeder of dogs or pigeons could perform nearly magical transformations over a few generations, then one could easily infer that animals in the wild could change indefinitely if they had endless ages to do it.

But even had Darwin been patient enough, he could not sit for ten thousand years and document the emergence of a longer beak, or the withering of an unused toe. He could only hope to find clues in nature that such changes had occurred. That meant looking for evidence in existing species branching off the evolutionary tree.

Branches, meanwhile, were still sprouting on Darwin's own tree. George Howard Darwin arrived at Down on July 9, 1845. Though never as famous as his father, George Darwin became an eminent scientist in his own right, as a mathematician and an astronomer. Two more Darwin children followed in short order: Elizabeth on July 8, 1847, and Francis on August 16, 1848, the same year that Charles's father died. Another son, Leonard Darwin, was born on January 15, 1850. He became the most prominent Darwin in the eugenics movement, drawing out more fully even than his father had that most basic tenet of Charles's theory: the more fit must replace the less fit. Leonard was followed by Horace Darwin on May 13, 1851. He became a maker of scientific instruments. His daughter, Ruth Frances Darwin, carried on one of the family traditions, becoming a passionate eugenicist along with her husband, Dr. William Rees-Thomas. Charles and Emma's tenth child, Charles Waring Darwin, arrived at the beginning of December 1856. The child was not quite right, and it is thought he may have been at least mildly retarded. But he was welcomed into a loving family. He died of scarlet fever in 1858.

Death and reproduction were, of course, central to Darwin's theory. He was fascinated by, of all things, the sexuality of barnacles. Sexuality was the source of nature's imprudent fecundity. Without it, there was no material for death to work on, to winnow, to select. And it was sexuality that allowed the selected traits to be passed on from one generation to the next. But how did sexuality itself arise? How did such a thing as man and woman, husband and wife, Charles and Emma, emerge?

The more immediately pressing issue, however, was how long the sickly and unfit Charles Darwin would live. His physical sufferings seemed almost to echo the trials of Job. His retching and reeling were amplified, and to these familiar maladies were added eczema, boils so painful he couldn't sit down to work, constipation, and most embarrassing of all, chronic flatulence that ensured he could spend very little time in polite company. Overwork was not the only culprit. Anxiety about death—the cold heart of his own theory—was another, as Charles fretted away during his father's declining health, and exploded in a shower of ugly symptoms after his passing.

He had thought being dashed with cold water brought on the symptoms of spiritualism; now Charles, completely desperate, latched onto the "water cure"—cold showers, steam baths, a strict diet (no beloved sweets, and soon enough, no dipping of snuff), walks in the fresh country air, and most of all, a moratorium on work. The "water cure" was all the rage among those who could pay to put themselves under a doctor's care at a lush spa, and although he was generally loathe to part with money, he was at wit's end. As with many quack cures, this one had an obvious grain of common sense encapsulated in nonsense.

Regularly dowsing Darwin with ice water could hardly have helped his condition, but the forced exile from work and the healthy food on offer at the spa did wonders for both his inner and outer maladies. So much so, that he installed the regimen at Down upon his triumphant return—including a strict limit on the hours he was permitted to close himself in his study. And the snuff? Well, Darwin tried manfully to shake his habit, but the habit was a more powerful friend than health.

While Darwin's health improved, three of his daughters were hit by scarlet fever in 1849, and one of them, Anne, was permanently weakened, developing similar problems with her stomach to those which Darwin himself was trying to shake off. Within a year, Anne's health began to worsen. Desperate—knowing he had bequeathed his own sickly constitution to his daughter—he took her to the spa at Malvern that had done such wonders for him. On April 23, 1851, she died. Darwin wrote back to his wife—who was at home, pregnant with Horace, and unable to travel—that Anne "went to her final sleep most tranquilly, most sweetly at 12 o'clock today. Our poor child has had a very short life but I trust happy, & God only knows what miseries might have been in store for her."[19]

Darwin's son Horace was born only a few weeks later and was a consolation, but both Emma and Charles were deeply wounded by Anne's death. Many biographers have portrayed Anne's death as marking a great theological turning point for Darwin, but they do so on the basis of sympathy, not evidence, for Darwin had already long lost his thin wisp of theism. His entire theory was designed, purposefully, to eliminate any need for God in the cosmos at all. If anything, one has to suspect that

Anne's death manifested itself to him as a painful, practical proof of his theory—death had reached into his family, into his heart, and wrenched his sickly child from his grasp. It was nature's way of removing the unfit and securing the life and inheritance of the fit.

In his despair he again immersed himself in his work, feverishly picking away, prodding, obsessed with his barnacles. He greedily dissected any specimens he could beg or borrow. He was now a renowned scientist, no longer climbing the ladder of the British scientific establishment, but sitting comfortably at the top. He was as humble and polite as ever, but he knew that when he sent a letter of request, the recipient would be eager to respond to the great Charles Darwin. Barnacles fascinated him, because they sorted themselves into distinct varieties, with distinct modes of reproduction, perhaps evolutionary modes: some appeared to reproduce as plants do, others as separate creatures in differing relations—a male "kept" by a dominant female as a kind of sexual prisoner, an independent male, and so on. Such, he crowed excitedly, must be the evolutionary story of the rise of sexuality, of Charles and Emma, of marriage, his own children, all romance and romantic poetry, love and love songs. He glimpsed the whole drama in what appeared to be the succession of evolutionary stages in the barnacles marching in front of his tired and triumphant eyes. His work on barnacles would win him a Royal Medal in 1853, putting yet another feather in his public cap, though he still breathed not a word that his research was intended to buttress an argument in favor of evolution.

The year of Anne's death and Horace's birth also marked the beginning of Darwin's friendship with Thomas Huxley, the man who rammed Darwinism broadside into the English scientific establishment. Without Huxley, it is safe to say there would have been no Darwinism. Huxley was like Grant, an iconoclast, a rebel, a man ready to tear down the established walls of Anglican state-church society. While Darwin quietly insisted on a godless account of the development of species, Huxley shouted down and humiliated any opponents of godless evolution. He didn't even agree with Darwin's account—maintaining, quite rationally, that the needed changes were too great to have been made at such an excruciating slow pace—but that didn't matter. Disagreements about the "mechanism" of evolution could be settled later. What mattered for Huxley now was that Darwinism could present a united front, or better a phalanx, to attack the privileged religious establishment, and free society for secularism.

At last, laying aside his writing on barnacles, Darwin took up the task again of filling out and firming up the large contours of his theory, but he did so not from philosophical reflection but further intensive study of particulars. He turned, in this case, to breeding pigeons and flowers, following up on his grandfather's assertion in the *Zoönomia* that just as man could shape species by breeding, so could nature do the same, guided by death weeding out the unfit—though of course whether death was actually comparable to intelligently directed breeding was, to say the least, an open question.

More than ever, his whole family became his staff, and Down his great laboratory. The entire family ordered its life around Darwin's work. Despite her misgivings at the direction, the purpose, of his research, Emma was always supportive of her husband. Without her scrupulous attention to his needs, he could never have done so much.

But as much as he did, and as much as was done for him, he could never feel himself ready. There were so many things he didn't know, so many objections that could be brought against his life's work. To expose it would be like exposing a weak but dearly loved child to assault, a child in whom one has every hope and expectation if only he could survive. Darwin needed encouragement. In the early spring of 1856, he invited his fervent allies Huxley and the botanist Joseph Hooker, as well as entomologist and malacologist (a specialist in mollusks) Thomas Wollaston, to Down for a full, unrestricted venting of his most radical ideas and their implications.

Ironically, Wollaston—whose own work, *The Variation of Species with a Special Reference to the Insecta; Followed by an Inquiry into the Nature of Genera*, was published in 1856—later became one of Darwin's sharpest critics, illustrating quite nicely the important distinction between Darwinism and evolution. He had no problem with the variation of species; his own fame resulted from his work on the variation of beetles. He had no problem with the fossil record. He had no problem with an earth much, much older than six thousand years. He was not a biblical literalist. He simply thought that Darwin had gone far beyond the evidence, and inferred far too much.

Earlier that year, Lyell had visited Down, and Darwin took him into his confidence about his theory of evolution. Lyell had the same reservations that others like Wollaston had, but he urged Darwin to publish anyway. The theory, as Charles spilled it out to him, was a worthy and powerful hypothesis, certainly far more worthy than had been made public by anyone else. Let it out in public, Lyell nudged, and see if it can stand the test of scientific criticism. For Darwin the urgency had another source. He feared being scooped again. "I rather hate the idea of writing for priority, yet I certainly shd be vexed if any one were to publish my doctrines before me."[20] *My* doctrines. My *doctrines*. A very curious and revealing way to put it.

He set to work on a massive account of his theory, one in which, as far as his strength would allow, he would fill the evolutionary framework with the endless details needed to vindicate it. He collected piles of his own evidence, and drew in, like a great vacuum, every conceivable scrap of information from scientists world-wide. He wanted his theory to be impregnable.

Who knows how long he would have worked on this massive volume—which surely would have stretched into a multivolume endeavor—if he hadn't received another nasty surprise in June 1858: a short essay that was a perfectly clear account of evolution by natural selection that couldn't have been a more accurate synopsis of his own, as if somebody had stolen the idea out of his head while he slept. It was written by Alfred Russel Wallace. Darwin's brainchild had been seized by another.

Wallace was halfway round the world. He had sent his short essay by post from the island of Temate in Indonesia. Darwin could have destroyed it, and hastily published his own account. No one would have been the wiser. But Darwin was a man of honor.

Still, all feelings of duty and honor aside, he felt mortally wounded, as if his life and life's work would expire, one with the other. To add insult to injury, Wallace had addressed it to

Darwin, hoping that he would pass it on to the more eminent Lyell.

But his wound aside, we again have to wonder at Darwin's surprise. To be so deeply immersed in his subject, and yet not recognize that others were following the some intellectual trends he was, and to the same conclusions—that is what is truly surprising. Certainly in retrospect we can say that if Darwin himself had suddenly expired, and all his notes had been burned, there can be absolutely no doubt that someone else would have taken Darwin's place in history. Perhaps it was that very unpleasant thought—that he was both unoriginal and intellectually expendable—that caused Darwin such anxiety. He thought he was harboring a great secret, which was in fact no secret at all.

After Darwin posted Wallace's package off to Lyell, he felt compelled to stake his claim as a theorist of evolution. Lyell, Hooker, and Darwin settled on a joint paper, hastily cobbled together from Wallace's essay and Darwin's private notes, a double statement of evolutionary theory, to be read before the Linnean Society on July 1, 1858. A few days before, however, on June 28, Darwin's young son Charles died of scarlet fever. Darwin could now not attend his own scientific coming out party. Instead, he sent his notes to Hooker to incorporate into the presentation. Wallace was still far away from England, and ignorant of the whole affair. The paper was read by a secretary of the Society, with Hooker and Lyell in attendance. Embedded among a long list of other papers, it received neither praise nor blame, but something more like fatigued indifference.

By August, Darwin was working on his own extended state-ment of evolution. No doubt he hoped this statement would eclipse the short paper presented to the Linnean Society earlier that summer. He would not raise his head until the task was fin-ished the following May. Emma helped in proofing the text. *On the Origin of Species by Means of Natural Selection, or the Preservation of Favoured Races in the Struggle for Life* appeared in print at the end of November 1859. Now the cat was fully out of the bag.

Chapter 5

One Long Argument, Two Long Books

It is customary to focus on Charles Darwin's *Origin of Species* as his *magnum opus*, the solitary peak of his evolutionary career, and then perhaps say a few things about his remaining works, including among them, almost as a casual aside, his *Descent of Man*. That is a mistake.

What we really have is one long argument in two long books, *The Origin of Species* and *The Descent of Man*, the first laying out his argument for evolution through natural selection, and illustrating it in regard to plants and animals; the second applying the theory to human beings. They should be inseparably bound together, for they only accidentally appeared in two parts, separated by a little over a decade. In reality, they had both been in Darwin's mind as one continuous argument for decades.

But at first Darwin only published half his argument, and this out of fear. First, he had been pushed into writing the *Origin* out of fear that someone else would beat him to it. Wallace's tightly wrought account of his theory finally made him realize that the same thoughts had occurred to others. We might find this astonishing since Wallace's essay is so short and Darwin's *Origin* is so long, but Darwin's book itself is not a complex, multifaceted, multilayered argument that slowly builds and reveals itself. It contains exactly one idea, which Darwin neatly states in the Introduction:

> As many more individuals of each species are born than can possibly survive; and as, consequently, there is a frequently recurring struggle for existence, it follows that any being, if it vary however slightly in any manner profitable to itself, under the complex and sometimes varying conditions of life, will have a better chance of surviving, and thus be *naturally selected*. From the strong principle of inheritance, any selected variety will tend to propagate its new and modified form.[1]

This one idea in and of itself was, however, entirely uncontroversial, *unless* it was taken to mean that forms, or species, might be modified without limit. And, of course, that is what Darwin sets out to convince the reader, using the Lyellian mode of approach to support endless slow modifications in biology. He did it beautifully, we should add; that is, the *Origin* certainly ranks as one of the greatest books of natural history on literary merit alone. Perhaps we have Emma to thank for some of that, but we add that Darwin's admiration for the Romantic naturalists had allowed the poet's eye for vivid and luxuriant detail to inform his vision of the drama of life.

Mellifluous prose would not be enough, however. Darwin knew the significant scientific arguments against unlimited transmutation. The fear of rebuttal had driven him to work diligently for nearly twenty years on the strongest possible counterarguments. His recognition that he had not overcome all these obstacles almost kept him from publishing, and when pushed to it, he manfully admitted in the very text of the *Origin* the great weak points, even indicating what evidence could fatally destroy his theory. But he was admirably tenacious and wonderfully clever in wringing every possible explanation supporting his theory out of an endless supply of facts.

There was also the fear of public recrimination of him as an atheist, a gutter radical, a destroyer of all things human. No one was fooled by Darwin holding off saying anything about human beings in the *Origin*. The monkeys to man bit had already been done by Lamarck and by Robert Chambers in his *Vestiges of Creation*. His own grandfather had been parodied as a transmutationist monkey by the Tory press some fifty years before the *Origin*. Everyone knew where the argument was headed, but Darwin thought if he left human beings out for the time being, then people might accept his argument for evolution, and he could slip man into the picture later.

Darwin was silent about God as well, but that silence was transparent in its implications: Darwin had not said anything about God because he had rendered Him entirely superfluous. He tried to patch things up in later editions, adding a sop for the incurably religious. The first edition ended with the famous flourish: "There is grandeur in this view of life, with its several powers, having been originally breathed into a few forms or into one...." To smooth ruffled feathers, later editions read: "There is grandeur in this view of life, with its several powers, having been originally breathed *by the Creator* into a few forms or into one...." Some are still fooled by this sop, but they shouldn't be, and most weren't at the time.

All of Darwin's fears, in fact, were realized upon publication. The shock to the public was not that someone had suddenly thrown a new and controversial theory in front of them. Even popular culture had been well aware of evolutionary arguments for at least half a century. But Charles Darwin was (unlike the anonymous author of *Vestiges*) a well-regarded scientist, honored by every scientific society, and respected by the public. He had published a string of highly detailed studies on coral reefs, volcanic islands, and such lovely creatures as the pedunculated cirripedes of Great Britain. He was riding high in the still largely Tory-dominated scientific establishment. Suddenly, with all his credentials and impeccable scientific reputation, he had become an intellectual turncoat.

Darwin was as shaken in the giving as the public was in the receiving. Unsurprisingly, the completion of the book in the spring of 1859 took place between retchings, and as soon as he had finished the manuscript, he hurried from Down to be doused with cold water and rubbed down, in an effort to resurrect his health. He had chosen another spa, Moor Park; he could not bring himself to go back to Malvern where he watched Anne breathe her last. He was entirely spent, his life's work draining the life out of him. He needed to think about anything other than the origin of species—a good smoke, long walks, a game of billiards, cheap romance novels, other literary works of higher merit (but not of an evolutionary sort), all these were salves for his mind.

While out walking one day, dutifully attempting to follow his doctor's orders to think of anything but evolution, he came across a line of ants, big red ants, carrying smaller black ones in their jaws. He could not resist. He had to watch. He needed help keeping track of particular ants, and dragged in a poor wandering vagrant who happened by, flipping him a shilling to add his eyes to Darwin's. Neither could blink or the particular ant each followed would be lost in a sea of *Formicae*. A carriage pulled by and just as curiously watched the two watchers. This became a family joke.

This comedy of curiosity contained a deep slash of ironic tragedy. As we have noted, Charles Darwin was a zealous abolitionist, the successor of a long line of passionate abolitionists. The Whig hatred of slavery thundered through Darwin's veins, and it bubbled to the surface at the mere mention of this foul institution. The slavery of one man by another was unjust and unnatural, a blight on human history that must be scrubbed clean. Whig history demanded its eradication as a necessary step of human progress. The Darwin-Wedgwood alliance was entirely of one mind and heart on this issue, and during the period before and after the publication of his *Origin*, Charles was grabbing every bit of news from America about the volcanic tremors of the slavery issue and the bloody Civil War.

But here he was, lost in thought and following the entirely *natural* enslavement of little black ants by big red ants. And it wasn't the first time. It was at Moor Park that Darwin had first witnessed natural slavery. He had before considered slavery as a human invention, and so he was astounded when he first witnessed for himself the "rare Slave-making ant." He crowed in a letter to Joseph Hooker that he had seen "the little black niggers in their master's nests."[2] A few months later, in July 1858, Darwin again reported jubilantly to Hooker that he was having "some fun here in watching a slave-making ant." The "wonderful stories" he'd heard about such ants—stories that sounded so much like the human slave trade—were true. Here in front of his eyes he watched "a defeated marauding party," and then "a migration from one nest to another of the slave-makers, carrying their slaves (who are *house* & not field niggers) in their mouths!"[3]

Darwin spent some time in his *Origin of Species* describing the "*slave-making instinct,*" detailing the exploits of the big red slavers, *Formica rufescens* and *Formica sanguine*, and the smaller black slaves, *Formica fusca*. Even in the more formal account in the *Origin*, he reported that he was initially skeptical that there might be a natural instinct for creatures to make slaves of others. Upon first hearing of such ants, he "tried to approach the subject in a skeptical frame of mind, as any one may well be excused for doubting the truth of so extraordinary and odious an instinct as that of making slaves." Yet, upon his own examination, he found "the wonderful instinct of making slaves" to be a natural fact, not a naturalist's fancy.[4] But if it is natural for ants, if it can be explained as the result of natural selection, then is it natural for human beings?

Darwin hated the thought. Quite obviously, he was of two minds about it, and that duality would prove to be the source of one of the deep tragedies inherent in his theory. Since he had set aside human beings from consideration in the *Origin*, he made no inferences about human slavery in the text. Yet, he could not avoid dealing with it forever. Perhaps his grandfather's abolitionist words, penned in his *Loves of the Plants*, were haunting him: "Hear this Truth Sublime, / He, who allows oppression, shares the crime."[5] Would his advocacy of evolution actually support slavery? Would he share the crime? How clearly did he even see the problem at this point?

When the *Origin* was finally published in November, Darwin was off at yet another spa for a cold baptism. He was miserably anxious now that the *Origin* had finally been born. There it was for all to see, and it was now time to orchestrate the reception of it. Darwin was a master conductor, pulling every string to get good reviews, calling in old cards, charming new prospective friends, especially useful, young, and energetic ones. He needed help trumpeting its strengths, but even more, in fending off attacks against its weaknesses. As he admitted privately to Harvard botanist Asa Gray, "There are very many difficulties not satisfactorily explained." But he held fast: since his theory explained a great deal, he reasoned, it couldn't be wrong. "On these grounds I drop my anchor, and believe that the difficulties will slowly disappear."[6] Asa Gray (along with Huxley, Hooker, and Lyell) became Darwin's intimate and active proponent, hatching strategies for dissemination of the *Origin*. Unlike Darwin, however, Gray thought that evolution could be reconciled with theism, for he was a firm believer.

As we might expect, his admirers played up the strengths while his detractors pounced on the weaknesses. His old friend and mentor Adam Sedgwick was shocked and crestfallen, and Richard Owen was outraged. Charles Lyell, who had helped guide the project through to its completion and had accepted a good bit of Darwin's theory, understood fully the moral and godless implications. He continued to stand by Darwin as a friend, but the counter-evidence bothered him. Most biographers paint Lyell as lacking courage and clinging to superstition, since he never went "full Darwin." Such is the continued influence of the Whig historians of science. The truth is otherwise. Lyell thought Darwin's theory was too small. It could not accommodate the real moral and intellectual differences that are the daily experience of a thinking, moral man.

Men like Lyell, and there were many, actually saw things much more clearly than did Darwin himself. Darwin always cheerfully assumed an evolution from savagery to civility. Since the brutality of natural selection had produced someone like himself—a gentleman, a very kind, honorable, affectionate, compassionate Englishman, and a liberal Whig to boot—he could safely assume a rosy secular future. But others, who very much appreciated Darwin's genius, were frightened by his blindness and the real heart of darkness in other men's souls. There was no reason why evolutionary competition should not yield something far more monstrous than Darwin. Darwin, they believed, entirely and naively mistook the nature of morality.

According to Darwin, morality does not govern evolution. If it did, then we might expect a divine overseer. Darwin would not allow that; and in order to disbar it, Darwin had to argue that morality was created by evolution. It is, in Darwin's scheme, an evolutionary after-effect of sociability. Natural selection first favored those human beings who were social, those that clung together in a group. Loners and the anti-social were unable to compete or simply killed as bothersome. The social instinct—which we share with a multitude of other animals from dogs to bees—became the foundation for other more complex habits that we now call "moral." Here is evolution in action:

> When two tribes of primeval man, living in the same country, came into competition, if the one tribe included...a greater number of courageous, sympathetic, and faithful members, who were always ready to warn each other of danger, to aid and defend each other, this tribe would without doubt succeed best and conquer the other.... A tribe possessing the above qualities in a high degree would spread and be victorious over other tribes; but in the course of time it would, judging from all past history, be in its turn overcome by some other and still more highly endowed tribe. Thus the social and moral qualities would tend slowly to advance and be diffused throughout the world.[7]

That doesn't sound so bad. Sure, survival of the fittest is a rather nasty road evolution has to take, but the end result is worth it—ever greater courage, sympathy, faithfulness. The future looks ever more moral.

> As man gradually advanced in intellectual power and was enabled to trace the more remote consequences of his actions; as he acquired sufficient knowledge to reject baneful customs

and superstitions; as he regarded more and more not only the welfare but the happiness of his fellow-men; as from habit, following on beneficial experience, instruction, and example, his sympathies became more tender and widely diffused, so as to extend to the men of all races, to the imbecile, the maimed, and other useless members of society, and finally to the lower animals—so would the standard of his morality rise higher and higher.[8]

But precisely here we run into trouble. There is no "higher" in evolution according to Darwin. Evolution doesn't aim at any-thing. It is governed by blind chance variations and ruthless cut-ting off of the unfit. If it did aim at some predetermined moral goal, one would immediately have to assume—as did Lyell, Gray, and Wallace—that a divine hand had stacked the deck. As Darwin himself defined it, *whatever* contributes to Tribe A's ability to survive at the expense of Tribe B, will define morality for Tribe A. It may be sympathy, but it also may be greater, more daring savagery. The same goes for the struggle of Race A against Race B, and Nation A against Nation B. Whatever it takes to win is by definition moral.

We can also see an interesting effect of this kind of argument. The moral traits Darwin finds congenial are the result of a long evolutionary contest, and precisely because he is ranking them, he must rank the human races according to their level of moral achievement. The result is that savages must be considered morally inferior in the same way that they have, say, a particu-lar shape of head or color hair. Nothing can lift them up the evo-lutionary ladder, morally or intellectually. This, in turn, though

Darwin did not want to think about this, provided a justification for slavery and the "slave instinct." The morally inferior races should be ruled for their own good.

But the deeper implications of Darwin's theory are, in fact, even worse. In Darwinian evolution there is really no such thing as morality at all. There are only moral*ities*, in the plural. Darwin insisted that morality itself must be reduced to a variable evolutionary phenomenon. Human moral*ities* are themselves the result of natural selection; that is, they are entirely defined by what qualities or traits have best enabled *particular* populations of human beings under *particular* circumstances to survive against other human populations. Moralities are evolved under quite particular historical and geographical contexts. There were endless varieties of them in the past and there will be endless possible variations in the future.

At Darwin's own insistence, nothing stands above these varieties of moral experience to judge them. The only criterion of judgment Darwin allowed is success in the struggle to survive. As with bird beaks and butterfly coloration, there are no right and wrong among moralities. There are good and bad, but not good and evil. But "good" means only whatever contributes to the survival of a particular people. That may mean increased sympathy toward their own tribe, race, or nation, but also increased savagery toward competing tribes, races, and nations. It could also mean the reverse, savagery among one's own tribe, race, and nation, if it contributed to the greater overall survival. It might mean the abolition of slavery, or the reinstitution of slavery. It could lead to vegetarianism or cannibalism. As many societies have amply proven, cannibalism isn't self-defeating:

you eat your enemies—natural selection and nourishment thereby go hand in hand.

And we must not forget that at the very heart, the very hard heart of Darwin's account of evolution, is death. Death stalks the weak, the sickly, the unfit, incessantly weeding them out, so that the strong may flourish. If the principle of natural selection completely replaces morality, then morality is reduced to natural selection. That is just what happened to Darwinism. It became the launching pad for the great eugenics movement, of which some of Darwin's sons and relatives became a part. Would Darwin have agreed with them?

You may judge for yourself. Here are the words of a man who, in writing the *Origin of Species* and the *Descent of Man*, was often too sick to work. Darwin matter-of-factly informs his reader that savages obey the laws of natural selection because they don't have the modern means to fight off the ravages of illness. As a consequence, "the weak in body or mind are soon eliminated; and those that survive commonly exhibit a vigorous state of health." But what about the civilized? According to Darwin,

We civilised men, on the other hand, do our utmost to check the process of elimination; we build asylums for the imbecile, the maimed, and the sick; we institute poor-laws; and our medical men exert their utmost skill to save the life of every one to the last moment. There is reason to believe that vaccination has preserved thousands, who from a weak constitution would formerly have succumbed to small-pox. Thus the weak members of civilised societies propagate their kind. No one who has attended to the breeding of domestic animals will doubt that this must be highly injurious to the race of man. It is surprising how soon a want of care, or care wrongly directed, leads to the degeneration of a domestic race; but excepting in the case of man himself, hardly any one is so ignorant as to allow his worst animals to breed.[9]

Sound like Ebenezer Scrooge? Or something worse, far more hard-hearted and hard-headed? But Darwin the man pulled back from the obvious implications of his theory, and took refuge in the evolved moral trait of "sympathy." We could not withhold "our sympathy, if so urged by hard reason, without deterioration in the noblest part of our nature," he maintained.

Hence we must bear without complaining the undoubtedly bad effects of the weak surviving and propagating their kind; but there appears to be at least one check in steady action, namely, the weaker and inferior members of society not marrying so freely as the sound; and this check might be indefinitely increased, though this is more to be hoped for than expected, by the weak in body or mind refraining from marriage.[10]

Darwin was so softened by sympathy that he could only bring himself to propose soft eugenics. His evolved trait of "sympathy" saved him from the more exacting demands of "hard reason." But this safe place of moral retreat was completely undermined by his own theory. Any evolved trait—including any evolved "moral" trait—exists in a particular individual or society *now* because it proved somehow beneficial to his

ancestors sometime way back *then*. Traits are considered
"noble" only because they have proven to be useful in the struggle to survive. But circumstances change, and hence natural
selection will choose other traits accordingly. There is no way to
predict what traits will be considered moral and noble. We will
know them when we see who comes out on top.

But even more damning for Darwin's attempt to escape from
his own conclusions is that sympathy for the weak, sickly, and
intellectually inferior is not just one more "moral" trait. It is a
trait that goes *directly against* the principle of natural selection.
Therefore, it does not take much imagination to see that Darwin's disciples would rightly insist that a society that goes
directly against the principle of natural selection will consign
itself to its own destruction—especially if it comes up against a
society that attempts to model its morality directly upon natural selection, a society that breeds for hard reason rather than
sympathy, and systematically destroys the weak, sickly, and
intellectually inferior.

In short, if the ruthless struggle to survive is the ultimate
cause of human morality, then nothing stands beyond evolution
to keep human morality from being entirely defined by that very
struggle. Such is the price Darwin paid for consistency, a Pyrrhic
victory, one might think, over the theistic objections of men like
Lyell, Gray, and Wallace.

Darwin genuinely believed his liberal Whig morality would
remain intact as evolution churned forward. That is one reason
he didn't shudder, like Lyell, at the entirely godless mechanism
of natural selection. In fact, he was very proud of it because it
so neatly eliminated the necessity for God. Yet, he was not

above using others, like the Reverend Charles Kingsley, who cheerfully accepted his theory and thought God could simply be put on top of it. When Kingsley praised the *Origin*, Darwin quickly put his words in the next edition, along with the sop about the Creator at the end. *He* didn't believe it, of course. The entire argument was designed to make God superfluous. But if folks like Kingsley thought they saw a place for God redundantly riding atop the whole thing, all the better for the ensuing propaganda campaign. It would help the acceptance of the theory, and having the theory accepted was vital to Darwin.

While Darwin gently assured sympathetic divines that there was nothing to worry about, Huxley harried them without mercy. Darwin believed in gradual transformation in intellectual, moral, and social life just as much as he did in geology and biology, which is why in good conscience he could accept theists who, in the course of time, he believed, would eventually dispense with their theism and embrace the full Darwin program. Darwin, then, was the tortoise. Huxley, who continually tweaked Darwin on the sluggishness of evolution, was the hare. He wanted both evolution and social revolution by leaps and bounds. He was a great believer in evolution precisely because it could bring about social revolution. He hated clerics with a Jacobin's passion, and he knew the acceptance of Darwin's theory meant the demise of the Anglican grip on English society. So, even though he differed with Darwin on how evolution actually worked—and much to Darwin's continual irritation pointed out the significant difficulties with natural selection in his public lectures—he bullied the opposition from his bully pulpit with evangelical zest.

One of Darwin's most zealous opponents was Bishop Samuel Wilberforce, the son of the great William Wilberforce who had spent his life working to eliminate the slave trade. The lives of the Wilberforces, Darwins, and Wedgwoods had been woven together in their opposition to slavery, and they still were. The good bishop was as vehemently abolitionist as his father. In the summer of 1858, a month before the Darwin-Wallace paper was read, Bishop Wilberforce was hectoring the House of Lords to force Spain to end the slave trade with Cuba. The heart of slavery was the "love of gain," a gain "purchased with...blood," and Britain is "bound by every obligation...most sacred" to put an end to such "evils."[11]

Bishop Wilberforce and Charles Darwin were of one heart about the evils of slavery, but not about evolution. Wilberforce attacked Darwin's theory with wit, reminding the public that the theory had descended with little modification from Erasmus Darwin; moreover, he made sure that the public was forewarned of its moral implications, namely that it would undo all the work the abolitionists had done to prove that slavery was immoral, wrong, and unjustifiable. If Darwin's theory were true, then human slavery was no less natural than ant slavery and hardly a matter for moral disapproval.

Our modern view of Bishop Wilberforce has been colored by the Whig version of history, of assured secular progress, which colors most modern textbooks. Thus Darwinian science is portrayed as triumphing over Christian superstition, with Wilberforce cast as a pompous and know-nothing clergyman clinging to his comforting myths rather than facing up to the scientific truth propounded by Darwin. In popular accounts of the great debate between Huxley and the bishop at the British Association for the Advancement of Science at the end of June 1860, we are led to believe that Bishop Wilberforce's religious bigotry was trounced by the effervescent Huxley. The alleged exchange goes something like this:

Bishop Wilberforce: "Are you related to an ape on your grandfather's or grandmother's side?"

Huxley: "I would rather have an ape for an ancestor, than use mere ridicule instead of rational argument. I choose the ape!"

And then, so it goes, the crowd erupted in cheers. All good men realized that the Church had been unmasked, and rushed to the side of Huxley and Darwin. Science rose victorious over biblical fundamentalism. Add to this picture the fact that FitzRoy, Darwin's old shipmate and captain, was in the audience, and publicly expressed regret at having ever been an accomplice to such an abominable theory. He waved the Bible in his hand, indicating his belief in the literal truth of Scripture and hence, a six-thousand-year-old Earth—a true Bible-thumper, further proof that opposition to Darwinism was limited to the stubbornly religious.

But this muddle of a mythical picture must be revised. First, while the Bishop was quite witty, as his father had been, he was also quite well read in science, and brought just the kinds of objections against Darwin's theory that other eminent scientists were offering. These were, of course, the very objections that Darwin feared, as they pinpointed the weak spots of his theory.

Second, and even more upsetting to the Whig version of the affair, Bishop Wilberforce was no unthinking biblical fundamentalist as he has been caricatured. His speech was based on a review of Darwin's *Origin*, in which he made it clear that Darwin's theory must be judged on the facts, and not on whether it seemed to contradict Revelation. He began,

> We are too loyal pupils of inductive philosophy to start back from any conclusion by reason of its strangeness. Newton's patient philosophy taught him to find in the falling apple the law which governs the silent movements of the stars in their courses; and if Mr Darwin can with the same correctness of reasoning demonstrate to us our fungular descent, we shall dismiss our pride, and avow, with the characteristic humility of philosophy... only we shall ask leave to scrutinise carefully every step of the argument which has such an ending, and demur if at any point of it we are invited to substitute unlimited hypothesis for patient observation, or the spasmodic fluttering flight of fancy for the severe conclusions to which logical accuracy of reasoning has led the way.

He ended,

> Our readers will not have failed to notice that we have objected to the views with which we are dealing solely on scientific grounds. We have done so from our fixed conviction that it is thus that the truth or falsehood of such arguments should be tried. We have no sympathy with those who object to any facts or alleged facts in nature, or to any inference log-

ically deduced from them, because they believe them to contradict what it appears to them is taught by Revelation. We
think that all such objections savour of a timidity which is
really inconsistent with a firm and well-intrusted faith.[12]

Finally, we must never forget the moral context of the
bishop's objections, and in particular, the issue of slavery. Huxley's witticisms deflect us from the fact that Wilberforce understood very clearly that Darwin's theory would undermine the
moral argument against slavery. He believed that evolution, particularly as Darwin had defined it, would destroy the arguments
of the abolitionists and reverse the forward campaign to eradi-
cate slavery entirely. This was a moral and scientific argument.
For Wilberforce, it was a *fact* that slavery was a moral evil. He
therefore warned Darwin after the *Origin* came out that it was
an unavoidable conclusion, according to his argument, that natural slavery among ants affirmed slavery as natural among men.
Wilberforce threw Darwin's own words back at him, that "the
tendency of the light-coloured races of mankind" to engage in
"the Negro slave-trade was really a remains of the 'extraordinary and odious instinct' which had possessed them before they
had been 'improved by natural selection' from Formica Poly-
erges into Homo."[13]
There were other facts brought to Darwin's attention by his
detractors, one being that evolution was old hat, while Darwin
had made it seem as if his theory were without precedent. At the
prodding of Asa Gray, he added a kind of overview to later editions of the *Origin* in which he acknowledged at least some of
his predecessors. Oddly, he thought that Aristotle had

foreshadowed him, rather than Lucretius (of whom he appears to have been ignorant).[14] Darwin cites a long list, including the Comte de Buffon, Lamarck, Geoffrey Saint-Hilaire, William C. Wells, the Reverend W. Herbert, Robert Grant, Patrick Matthew, Leopold von Buch, Constantine Samuel Rafinesque-Schmaltz, Jean Baptiste Julien d'Omalius d'Halloy, Wallace, Henry Freke, Herbert Spencer, Charles Victor Naudin, Hermann Schaaffhausen, Thomas Huxley, and Joseph Hooker bringing up the end. He even mentions Erasmus Darwin in a footnote. All had, in one form or another, affirmed some form of evolution, some kind of descent with modification, some essential aspect of natural selection. It was not *his* theory.

The years following the first appearance of the *Origin* were marked with significant success in getting the message out. As the list of predecessors makes evident, the modification of species was already taken seriously by a wide number of scientists and allied intellectuals. Darwin was not alone in pointing this out. The notion that species changed over time, at least to a limited extent, was already well on its way to acceptance even without the *Origin*, and Darwin's cogently reasoned arguments and ample detail won even more converts. But the victory of Darwinism was not by reason alone. It was the result of a well-planned intellectual and social revolution. As biographer Janet Browne notes, in the decades following the publication of the *Origin,*

Darwin's defenders came to occupy influential niches in British and American intellectual life. Together, these men would also control the scientific media of the day, especially the important journals, and channel their other writings through a series of carefully chosen publishers—Murray, Macmillan, Youmans, and Appleton. Towards the end they were everywhere, in the Houses of Parliament, the Anglican Church, the universities, government offices, colonial service, the aristocracy, the navy, the law, and medical practice; in Britain and overseas. As a group that worked as a group, they were impressive. Their ascendancy proved decisive, both for themselves and for Darwin.[15]

The strategy was largely defined very early by that small band of revolutionaries who were determined to make Darwin's theory orthodoxy—Darwin himself, Hooker, Lyell, Huxley, and Gray. Huxley and several other devotees even started the "X Club," something like Erasmus Darwin's Lunarians, but devoted entirely to the task of dissemination of evolutionary theory. The strategy paid off as the circle widened, and soon pressure could be exerted from the top down, once scientific societies, publishers, universities, and journals were commandeered. (In fact, the prestigious journal *Nature* was founded in 1869 as an organ for disseminating Darwin's thought.) Opponents were locked out, ignored, and mocked. As Browne reveals, Darwin made his contributions from behind the scenes, letting his more forceful proponents do the direct work of takeover. And such efforts were not confined to Britain. Darwin used Asa Gray at Harvard to help take America, and Darwin himself saw to the translation of the *Origin* into German and French, and urged his contacts on the continent to help spread his theory.

All this work on behalf of the *Origin* was exhausting to Darwin—and he was still a family man. During this period, his daughter Henrietta began to show signs that she had inherited his weak constitution, especially his wayward stomach. Having lost one daughter, his beloved Anne, every bad report on Henrietta's health sent Darwin into stomach spasms of his own. Poor Emma had to act as nurse to both. Henrietta would recover, but Darwin's health only seemed to get worse, the least little exertion throwing him into a round of retching and bed rest that now lasted for days rather than hours. He was turning into an old man, an invalid. Vomiting, dizziness, chronic flatulence, eczema—every effort on behalf of his work seemed to stir the ever-present murmur of ill health into full song. The youngest child, Horace, appeared also to have inherited his father's weak stomach, a fate which Darwin cursed all the more as he realized he himself had been its source.

Yet Darwin persevered in his work. The Darwin household at Down would become, more and more, one great experiment, with all the children and servants cheerfully corralled into helping him watch over the smallest details of life to see if he could catch glimpses of evidence that would support his argument.

Darwin was absorbed during the years after the *Origin* in teasing out the details of the sexual life of orchids, and the extraordinary dividing-line creature, the carnivorous plant the sundew. This was, of course, the kind of research that Grant had put him up to so many years ago at Edinburgh: look at the fuzzy lines between plants and animals. Whatever seems to sit there will reveal clues that will show that no line exists at all, and evolution has therefore easily climbed the ladder between them.

While sickness was Darwin's close companion, death visited his beloved friends during these years. Huxley's son, Noel, was taken the September after the great debate with Bishop Wilber-force, but Huxley would not flinch. "I could have fancied a devil scoffing at me," he wrote, "and asking me what profit it was to have stripped myself of the hopes and consolations of the mass of mankind? To which my only reply was & is Oh devil! truth is better than much profit."[16] For all his bravado, Huxley was so grief-stricken as to be severely imbalanced, and Darwin worried about him.

The following May, Darwin's dear friend, mentor, and benefactor John Henslow died. Darwin could not bring himself to visit him while he was dying. The anxiety caused him nearly a day's worth of retching, or so he said, though he seemed well enough to be trotting around London. Still, we cannot be too hard on Darwin. I think he genuinely loved Henslow too much to watch him sink away before his eyes. Henslow was gracious to Charles right to the end, urging that his theory should be taken with the utmost seriousness, even though he himself found it ultimately unconvincing.

The great historical irony is that even among his closest allies—Hooker, Huxley, Gray, Lyell, and Wallace—there were doubts about Darwin's theory. It is precisely here, among his allies, that we find the most interesting evidence to contradict the notion that all good, intelligent, and honest men leapt immediately to the conclusion that Darwin was completely right. Even they had significant misgivings, and their hesitations grieved Darwin.

Asa Gray was Darwin's man in America, helping him to thump his fellow Harvard luminary, Louis Agassiz, who not only rejected Darwin, but affirmed a pro-slavery account of the different races as separate and unequal creations. Gray wrote a three-part defense of Darwin for the *Atlantic Monthly*, a kind of theistic justification of the arguments of the *Origin*, with the drawn-out reprinted title *Natural Selection not inconsistent with Natural Theology: A Free Examination of Darwin's Treatise on the Origin of Species and of its American Reviewers* (1861). Darwin was delighted with it, and had paid for the cost to reprint it, sending it off to strategic places to blunt the theological objections and recriminations. Quite unlike Darwin, Gray believed in God, and saw no difficulty in His having shifted a great part of the work of creation to secondary causes, to the creatures themselves. Contrary to Darwin's own intentions, Gray asserted that his theory did not eliminate divine design, but lifted it to a new, more intricate level. And unlike Darwin, Gray believed that the human mind could not be explained as the material result of natural selection—Gray himself probably did not realize at the time that his view differed from Darwin's on this point, as he told readers that he presumed Darwin would not accept the development of mind from instinct.[17] He provided what amounted to a rather ingenious revitalization of natural theology, more sophisticated even than William Paley's.

The problem was, of course, that Darwin himself had designed the theory to eliminate any connection to God whatsoever. He disagreed with Gray's theological spin entirely, and was perhaps peeved by some of Gray's implicit criticisms of his atheism, and the materialistic foundation of his argument. That is not what he *meant* the theory to do, and in private letters he politely made his objections known to Gray. Yet—and this was typical of Darwin—he had no qualms about using Gray's argument if it would smooth the way for acceptance of his theory. Once the theory was accepted, the theistic patina would be ground away by the hard, anti-theistic core of the argument.[18]

Lyell both supported and disappointed Darwin as well. He had given up his long-term animosity to transmutationism, and in fact, published a very influential book in 1863 that searched where Darwin had not dared to go in his *Origin*—the dark path of human origins: *The Geological Evidences of the Antiquity of Man with Remarks on Theories of the Origin of Species by Variation*. Lyell was a great man of science and an irreplaceable ally to Darwin. Darwin passionately wanted him to affirm the complete evolutionary continuity between the higher apes and human beings. Lyell's prestige would make it that much easier for Darwin to publish what he had been holding back. But for Lyell, the evidence of man's singular moral and intellectual abilities that set him *apart* from the merely natural evolutionary continuum, was just as real and solid as the evidence for the natural evolutionary continuum itself. Darwin was "deeply disappointed" by Lyell, and he let him know it.[19]

And then there was Wallace, whom Darwin himself was compelled to affirm as the co-discoverer of the theory of evolution through natural selection. Wallace had originally been quite satisfied to derive the entire range of human capabilities from natural selection working on random variations, but the more he thought about it, the less plausible it became. In a paper of 1869 dealing with Lyell's geology and the *Origin of Species*, Wallace politely parted company with Darwin. Affirming natural selection for the entire plant and animal kingdom, he argued that the "intellectual capacities and his moral nature were not wholly developed by the same process." He wrote further:

> Neither natural selection nor the more general theory of evolution can give any account whatever of the origin of sensational or conscious life. They may teach us how, by chemical, electrical, or higher natural laws, the organized body can be built up, can grow, can reproduce its like; but those laws and that growth cannot even be conceived as endowing the newly-arranged atoms with consciousness. But the moral and higher intellectual nature of man is as unique a phenomenon as was conscious life on its first appearance in the world, and the one is almost as difficult to conceive as originating by any law of evolution as the other. We may even go further, and maintain that there are certain purely physical characteristics of the human race which are not explicable on the theory of variation and survival of the fittest.[20]

For Wallace, one of the most telling problems with pure Darwinism is that the lowest "savages" have moral and intellectual capacities that could not possibly have been derived by selection in the primitive circumstances. "The higher moral faculties and those of pure intellect and refined emotion are useless to them, are rarely if ever manifested, and have no relation to their wants, desires, or well-being." Rather than seeing in (say) the Fuegians the low end of a smooth intellectual and moral continuum down to the apes, Wallace argued that there was a huge evidential gap. "How, then, was an organ" like the brain "developed so far beyond the needs of its possessor? Natural selection could only have endowed the savage with a brain a little superior to that of an ape, whereas he actually possesses one but very little inferior to that of the average members of our learned societies." But the same goes for the hand, the upright posture, the organs of speech—in each case the "savage," who has no need of their great intricacy, differs inconsequentially from the most advanced European.[21] That would obviously make a very good argument *against* slavery, even as Darwin's theory seemed to set a natural precedent for it.

Wallace concluded with an invitation to consider amending Darwin's account. We must open ourselves to "the possibility of a new stand-point for those who cannot accept the theory of evolution as expressing the whole truth in regard to the origin of man. While admitting to the full extent the agency of the same great laws of organic development in the origin of the human race as in the origin of all organized beings, there yet seems to be evidence of a Power which has guided the action of those laws in definite directions and for special ends." We may then look forward to a "true reconciliation of Science with Theology." The final words had the tone of an exhortation: "Let us not shut our eyes to the evidence that an Overruling Intelligence has watched over the action of those laws, so directing variations and so determining their accumulation, as finally to produce an organization sufficiently perfect to admit of, and even to aid in, the indefinite advancement of our mental and moral nature."[22]

Darwin was absolutely livid when he read Wallace's piece. "No!!!" he scribbled in the margin. "I hope you have not murdered too completely your own & my child . . . I differ grievously

[sic] from you, and I am very sorry for it." He turned to Lyell for comfort, but Lyell only lauded Wallace's emendation of Darwin's theory.[23]

Opposition from opponents was to be expected. Opposition from one's intimate friends was another thing altogether. Darwin would not let his "child" fall into the hands of others. Sometime around this period, Darwin decided that he must speak out. He had to show how his theory could indeed smooth out the mental and moral differences between animals and human beings. The result was the second part of his one, long argument, *The Descent of Man, and Selection in Relation to Sex*, published in 1871. His daughter Henrietta proved an invaluable editorial aid, and his wife, Emma, read the manuscript as well, patiently helping her husband, even as she agonized over the state of his soul and the eternal distance he was determined to create between them.

In the *Descent* Darwin marshaled every scrap of evidence for his case that the argument of the *Origin of Species* could be directly applied to man without remainder. However detailed the scraps, the basic idea was quite simple, and he simply applied that one idea to man. If we were to alter his words from the *Origin*, it would go like this:

As many more individuals of the *human* species are born than can possibly survive; and as, consequently, there is a frequently recurring struggle for existence, it follows that any *human* being, if it vary however slightly in any manner profitable to itself, under the complex and sometimes varying conditions of life, will have a better chance of surviving, and thus be *naturally selected*. From the strong principle of inheritance, any selected variety *of human beings* will tend to propagate its new and modified form.

We must remind ourselves that the real opponents to this continuation of his theory were those men whom Darwin most wanted to convert, his friends Lyell, Gray, and Wallace. If his supporters jumped ship on this most important of points, then who would support him? Allowing a moral and intellectual leap for humanity—a leap that natural selection could not make—would be tantamount to reaffirming some form of theism. For Darwin, this was unconscionable.

Darwin used a simple and brilliant two-pronged strategy. First, he must show that the rudiments of moral and mental capacities can be found in other animals, and that human beings *themselves* manifest a kind of graded intellectual and moral hierarchy. If such is the case, then an evolutionary continuum is certainly plausible. But plausibility based on similarity was not enough, as Darwin realized. And so, secondly, he had to give an account of how our moral and intellectual faculties had been derived through natural selection. As he carried out this strategy with admirable thoroughness, he revealed the dark side of Darwinism, the side that Darwin himself could not fully face, although he did not flinch until the last moments.

As to the first prong, Darwin laid out his case that other animals show clear evidence of rudimentary moral and intellectual capacities. The "lower animals, like man, manifestly feel pleasure and pain, happiness and misery." They express "Suspicion . . . Courage and timidity," and in dogs and horses we find some "are ill-tempered and easily turn sulky; others are good-tempered; and these qualities are certainly inherited." We find "maternal affection," jealousy in both dogs and monkeys, and "all animals feel Wonder, and many exhibit curiosity." Finally, "Few persons any longer dispute that animals possess some power of reasoning. Animals may constantly be seen to pause, deliberate, and resolve." Furthermore, they have some "idea of property" in regard to defending their territory; apes use rudimentary tools and build shelters; and, from bird songs to dog barks, we find rudimentary language. The combination of rudimentary language and increased brain power "in some early progenitor of man" was the evolutionary avenue for our own far more developed powers. The increase of vocabulary itself can be put down to evolution, "for there is in the mind of man a strong love for slight changes in all things. The survival or preservation of certain favoured words in the struggle for existence is natural selection." Art is no exception, since animals have a sense of beauty, as the elaborate plumage of male birds seems to show. [24] And religion? First of all, Darwin wrote, "There is no evidence that man was aboriginally endowed with the ennobling belief in the existence of an Omnipotent God. On the contrary... numerous races have existed and still exist, who have no idea of one or more gods, and who have no words in their languages to express such an idea." The belief in "unseen or spiritual agencies," however, "seems to be almost universal with the less civilised races," but that proves nothing, since savages are likely to project the figures in their dreams onto reality. Moreover, Darwin's own dog illustrates quite clearly the origin and caliber of religious belief. He was lying in the sun one day, and an umbrella nearby kept being stirred by the breeze. Every time the umbrella moved, "the dog growled fiercely and barked. He must, I think, have reasoned to himself in a rapid and unconscious manner, that movement without any apparent cause indicated the presence of some strange living agent. . . ." From just such beginnings, the "belief in spiritual agencies would easily pass into the belief in the existence of one or more gods. For savages would naturally attribute to spirits the same passions, the same love of vengeance or simplest form of justice, and the same affections which they themselves experienced." The feeling of awe and wonder? Well, "a dog looks on his master as on a god."[25] In sum, there is no leap, but a moral and mental continuum.

Quite logically, but insidiously, the second prong is implicit in the first. The great leap between the highest animal and the highest human being—between, say, an orangutan and Darwin himself—must itself be closed by a series of steps from the lowest savage with the least bit of intellectual advance over the orangutan to the man capable of setting out the theory. "Differences of this kind"—moral and intellectual—"between the highest men of the highest races and the lowest savages, are connected by the finest gradations."[26] We repeat his assessment of the Fuegians, that "in this extreme part of South America, man exists in a lower state of improvement than in any other part of the world."[27] They exist as evidence for his overall argument. Human beings have, Darwin asserts, a common ancestor; they "are descended from a hairy quadruped, furnished with a tail and pointed ears, probably arboreal in its habits. . . . But since he attained to the rank of manhood, he has diverged into distinct races, or as they may be more appropriately called, sub-species." So Darwin could write: "Some of these, for instance the Negro and European, are so distinct that, if specimens had been brought to a naturalist without any further information, they would undoubtedly have been considered by him as good and true species." Because they have a common progenitor, all the races "agree in so many unimportant details of structure and in so many mental peculiarities," and that progenitor "would probably have deserved to rank as man."[28] Yet, since evolution has continued to work, the effects of natural selection in pick-ing some "favoured races" to climb higher is expressed in the differences between the races themselves; that is, the existing races themselves can clearly be ranked from lowest to highest, the lowest being most similar to the common progenitor (and hence, most similar to the highest existing apes) in moral and mental abilities. The Fuegians were not too far above the apes.

Where, we might wonder, was Darwin's heart? His head was quite logical, following an impeccable line of reasoning from his first premises to the full conclusions. But as Bishop Wilberforce had duly warned him, these conclusions supplied exactly the kind of scientific foundation that pro-slavery racists wanted. In fact, he gave them *more* than they dared dream of asking.

Evolution climbs by extinction in the struggle for existence, individual against individual, species against species, tribe against tribe, race against race. It was an essential part of Dar-win's argument that, in the struggle to survive, "the forms which stand in closest competition with those undergoing modification and improvement will naturally suffer most." The "most closely-allied forms" struggle over the same turf, and those who are better equipped through random variation will win out over the less endowed, whether it is a slightly longer beak for a finch or a bit more intelligence or courage for a particular human tribe or race. "Extinction follows chiefly from the competition of tribe with tribe, and race with race." The fit survive; the unfit are destroyed. How do we *know* that the victors are really more fit? By the very fact that they annihilated another tribe or race. That is the way of evolution. To condemn the slaughter of one tribe or race by another is to condemn evolution itself. That's how we got where we are today.

But as Darwin makes stunningly clear, evolution does not stop, even for human beings. And so, he offers the following conclusion the context makes it all the more interesting. The context is provided by a most serious objection to Darwin's theory, one his own friends pointed out. What about the great gap that exists between the highest apes and man? Darwin replies, "The great break in the organic chain between man and his nearest allies"—the so-called missing links—is not some kind of singular, startling anomaly. Such breaks or gaps occur everywhere in evolution precisely because the things that are closest to each other (the "most closely-allied forms") kill each other first, thereby creating the "gap." The same thing happened with man as he climbed up from a common ancestor. The apes who were a little more like man killed those a little less like man, but didn't exterminate those who were a lot less like men; the men who were a little better endowed with reason and fighting ability, killed those men who were less well-endowed, but left the apes who were farther down the scale. Hence, the seeming breaks between species. The same process *must* continue, and the gaps will therefore only widen. Here is Darwin's horrifying finale.

> At some future period, not very distant as measured by centuries, the civilised races of man will almost certainly exterminate and replace throughout the world the savage races. At the same time the anthropomorphous apes [like the gorilla, orangutan, or chimpanzee] ...will no doubt be exterminated. The break will then be rendered wider, for it will intervene between man in a more civilised state, as we may hope, than the Caucasian, and some ape as low as a baboon, instead of as at present between the negro or Australian and the gorilla.[29]

I think it is fair to say that, at this point, Darwin had contradicted himself. By this, I do not mean that his theory had a contradiction. Quite obviously, it has a dreadful and ruthless consistency. I mean that Darwin's theory, which he believed in with all the passion of a religion, which he cared for with all the fierce tenderness of a father for his beloved child, contradicted Darwin's own most sincere humanity, his hatred of slavery, his native abhorrence at seeing anyone or anything suffer, his almost epic kindness and gentleness with his family, friends, servants, neighbors, and strangers. That is a puzzle about which we should be deeply worried, since the dreadful and ruthless consistency would be carried forward by Darwin's disciples, who admired Darwin's head but lacked his heart.

Speaking of heads, for Darwin as an evolutionist who reduced the mind to matter, the shape and size of the human head was the single most important clue to the slow transformation of ape to human, and then from the lowest to the highest human types. As he argued in his *Descent of Man*, "There exists in man some close relation between the size of the brain and the development of the intellectual faculties," a fact that is "supported by the comparison of the skulls of savage and civilized races, of ancient and modern people, and by the analogy of the whole vertebrate series." Even today, the "mean capacity of the skull in Europeans is 92.3 cubic inches; in Americans 87.5; in Asiatics 87.1; and in Australians only 81.9."[30]

No one who has viewed the later pictures of Darwin, which display his prominent bald head, so large that it looks like he's wearing a football helmet, can doubt where Darwin thought he stood in the evolutionary hierarchy. That we find so many pictures of Darwin after the *Beagle* years with his hat off makes us suspect him of a kind of peacock pride in displaying his ponderous pate. But we must never forget that foolishness had serious moral implications: the size of the head was further material evidence of the ranking of the races, and hence which ones could safely be jettisoned as intermediate species in the ongoing struggle of human evolution.

For all these reasons, we must also reject the caricature of Darwin as representing the righteous struggle of science against dread superstition, the very Whig view of history that so thoroughly informed three generations of Darwins, the view into which Darwin saw himself fit. First of all, his theory brings no march of moral progress, of righteousness. Despite his intentions, Darwinism leads to the defeat of morality and the victory of a sophisticated kind of barbarism. Second, his theory certainly appears to be ideologically bent, rather than scientifically straight. It may just be the case—and I think it is—that Darwin's theory of evolution *itself* was formed to fit into the Whig history of materialist science triumphing over spiritualist superstition. That is why he felt he had to make it godless. Larger theories were possible, and in fact, demanded by the evidence. Materialist reductionism was not the only possible approach.

I think that is precisely what Emma, his wife, was trying to tell him. She was not afraid of him being a scientist. If he were pursuing science like Sedgwick, Henslow, Owen, or even Lyell, Emma would have had little to fear. They did not consider belief in God to be systematically antithetical to science. But something about the *way* Darwin approached things led him to define science *against* belief in God.

Emma, who made no pretensions to intellectual expertise in philosophy, issued a warning to her husband that was rediscovered as a profound truth over a century afterwards by philosophers and historians of science such as Michael Polanyi and Thomas Kuhn.[31] "May not the habit in scientific pursuits of believing nothing till it is proved," asked Emma quite reasonably, "influence your mind too much in other things which cannot be proved in the same way, & which if true are likely to be above our comprehension." The chosen scientific hypothesis or paradigm, the lens through which the investigator attempts to scrutinize nature, both magnifies and distorts, bringing objects nearer and crowding them within a particular field of vision, but at the expense of what lies outside and beyond the frame.

Emma was asking him a question both simple and insightful. Could the wisdom of God, God Himself and His mysterious ways, be too large and untamed to fit within Darwin's conception of what counted as demonstrable? That is a most profound question. It is the exact same kind of question as the following. Could nature be *more* elaborate, *more* wonderful and strange, than Darwin imagined it? Might evolution itself be a process far more intricate, grand, and mysterious than could be neatly reduced within the tightly sealed confines of natural selection? Can the expansive mind of man really fit the contracted view of human reason Darwin needed to fit his materialism? Or, to put it most pointedly, is Darwinism large enough to explain Darwin himself, a man of evident genius?

Chapter 6

Darwin Meets His Maker

When Darwin published the *Descent of Man* he was not much beyond sixty years old, but he felt and looked much older. His continual wrangling with his stomach, head, and skin had drained years from his life. He was a frail old man. In one sense, he had nothing left to do but die. He had said everything he needed to say, laid out the whole evolutionary framework from beginning to end. But in another sense, he knew he had work to do, work that he could not possibly live to finish. As he himself admitted, the argument was hypothetical. He had given a persuasive argument. The vast framework was there, but the details needed to be filled in, and the details were as endless as the intricacies of nature itself. Even worse, the legitimate objections kept arising like the heads of the hydra. With every objection he answered, another two would take its place.

A particular thorn in his side was a former pupil of Thomas Huxley himself, the English anatomist and biologist St. George Jackson Mivart. He had begun, under Huxley's tutelage, as an ardent evolutionist. He remained an ardent evolutionist, and indeed put forth a form of theistic evolution not too different from Darwin's co-discoverer, Alfred Wallace. Wallace looked at the evidence of humanity's moral and intellectual capacities, and reformed Darwin's theory to accommodate a divine hand in bringing about the great leap to human evolution. Wallace, however, was not an orthodox Christian believer. Guided by scientific and agnostic premises, he ended up in a vain search among spiritualists. It was a path that many other Englishman who surrendered their Christian faith were to follow. Spiritualism, like phrenology, was a repository for people who thought themselves more scientific and rational than your average superstitious Christian, and so thought they were called upon to improve upon revelation. Wallace remained Darwin's friend, even though Darwin groaned at Wallace's attempts to redefine his theory.

St. George Mivart, on the other hand, was a convert to Catholicism who saw the same evidence of a great moral and intellectual leap from ape to man. He tried to reconcile evolution with Christian orthodoxy, and did so by hammering away at the weak points in Darwin's theory. The weakest of all, in Mivart's opinion, was that it was undergirded by naturalistic philosophical presuppositions that were unexamined. Darwin had rejected outright the idea of a divine hand guiding evolution, and then set about explaining religion away as one more unintended effect of natural selection. To do so, he had to treat theology in a most superficial and unsophisticated way. Darwin's theory did not prove that there was no Creator God; it began from the assumption that God did not exist, and so his theory was constructed and expressed in such a way as to dismiss the possibility without seriously engaging it. Moreover, in order to ensure that a divine foot could not enter the door, Darwin had skirted over the profound moral and intellectual differences between man and animals by pretending that an enormous gap didn't exist when in fact it most obviously did.

Darwin could ignore many of his critics, but Mivart's arguments and credentials were formidable. A professor of biology; Fellow and Vice-President of the Zoological Society; Fellow, Secretary, and Vice President of the Linnean Society; Fellow of the Royal Society; he would later get his doctorate in medicine from the University of Louvain. Almost simultaneously with the publication of Darwin's *Descent of Man*, Mivart came out with *On the Genesis of Species* in which Darwin's *Origin of Species* (including its implications for man) was subjected to the most thorough and serious criticism in Darwin's lifetime. Mivart's attack was so devastating that Darwin felt as though he had to begin anew.

Again, the problem wasn't evolution. Mivart believed that the current understanding of evolution was still largely hypothetical, but that the facts would, more and more, support it, and further, that evolution would prove itself to be "perfectly consistent with the strictest and most orthodox Christian theology."[1] But Darwin's version was another thing: "The special Darwinian hypothesis, however, is beset with certain scientific difficulties, which must by no means be ignored, and some of which, I venture to think, are absolutely insuperable."[2] Two related problems must be mentioned. First, natural selection itself is insufficient to explain the evolution of species and so must "be supplemented by the action of some other natural law or laws as yet undiscovered," and second, illegitimate consequences have been drawn from evolution against religion.[3] Notice the relationship between the two. Darwin had carefully crafted natural selection to displace the need for God, but natural selection itself was inadequate on *scientific* grounds. It was hamstrung by its own reductionist, anti-theistic bias, the very bias which led to its illegitimate attack on religion as a mere epiphenomenon of natural selection.

What were some of the inadequacies of natural selection? Natural selection is "incompetent to account for the incipient stages of useful structures" because these first stages themselves cannot yet contribute to survival (and so, wouldn't be selected); similar biological structures develop from wholly different origins, something that couldn't happen by mere random variation; there are biological grounds for believing that the evolutionary transition between species "may be developed suddenly instead of gradually" (as Darwinian gradual transformation demanded); as Lyell and others had pointed out, "species have definite though very different limits to their variability"; "certain fossil transitional forms are absent, which might have been expected to be present" if, as Darwin maintained, evolution without divine help always had to move ahead by the tiniest steps; and finally, "there are many remarkable phenomena in organic forms upon which 'Natural Selection' throws no light whatever.... "[4] As to the last, an interesting example brought forward by Mivart was the number of fish, "such as the sole, the flounder, the brill, the turbot," that when young have eyes on both sides of their heads and swim upright just as other fish, but then as adults their two eyes have moved to one side of their heads so that they can swim on their sides, flattened on the bottom of the sea or ocean. It is certainly quite useful to be a fish that can flatten itself upon the sea bottom, but "how such transit of one eye a minute fraction of the journey towards the other side of the head could benefit the individual," and each such fraction be chosen as contributing to survival, "is indeed far from clear."[5]

Darwin took up the *Origin*, which was going into its sixth edition, and added an extended reply to Mivart. It was Darwin's longest rebuttal against any opponent, a sign of how deeply Mivart had gotten under his skin. Unlike the belligerent Huxley, Darwin was always the gentleman, though he did quietly cut Mivart from his list of allies and friends. "A distinguished zoologist, Mr. St. George Mivart, has recently collected all the objections which have ever been advanced by myself and others against the theory of natural selection, as propounded by Mr. Wallace and myself, and has illustrated them with admirable art and force."[6] Yet, Darwin did not back down. Sometimes he rises to the occasion and gives a truly plausible answer; other times his ingenuity fails him. Mivart had asserted that, if natural selection was always pushing creatures to maximize their survival abilities, then one would expect more animals than just the giraffe to have developed long necks to browse on the upper reaches of trees during a drought.[7] Darwin answered that, as can be seen in the English meadows, cattle and sheep graze at different heights, and so on the savannah, other animals would benefit by grazing at lower heights, so as not to compete with the giraffe.[8] However, on flat fish Darwin fell flat, and fell into Lamarckism, asserting that young flounder (for example) fall over sideways, due to their bodyweight distribution. They then strain their bottom eye in an attempt to look upward. Over time this would lead to an inherited malformation in the skull and a wandering eye.[9] But why, one might legitimately ask, would a fish with such an unhelpful bodyweight distribution ever have survived in the first place? Why would natural selection have chosen increments leading up to a manifestly ill-fitted trait, that caused the fish to flop over on its side so that it had to strain to see upwards so hard that it slowly malformed its skull? And why would a minute malformation of the skull, caused by an eye still peering at the sand, bring any benefit whatsoever? Why and how would any of this be inherited? The list goes on.

The "admirable art and force" of the arguments Darwin felt compelled to answer came from the fact that Mivart was himself an evolutionist, a distinguished scientist who had every right to claim to be at least Darwin's equal if not superior, and had philosophical and theological training that far exceeded Darwin's. Mivart gladly accepted the strong points of Darwin's argument, but accused Darwin of sowing conflict between evolution and the idea of a Creator God where none existed. Mivart found no such conflict and wrote that "Christian thinkers are perfectly free to accept the general evolution theory." [10] In fact, Mivart posited, evolution itself *needed* a divine explanation. Again, the very faults of Darwin's account which Mivart so deftly pegged were not problems with evolution per se, but were caused by Darwin's insistence on a purely reductionist, materialistic account. But that was a *philosophical* problem in Darwin, not a scientific one, and Mivart went on to show how two great doctors of the Church, St. Augustine and St. Thomas, both insisted that God regularly worked through secondary causes in nature (St. Augustine had even at one time outlined something of an evolutionary theory himself). Theism was not, as such, opposed to evolution properly understood. Mivart concluded,

> The Author ventures to hope that this treatise may not be deemed useless, but have contributed, however slightly, towards clearing the way for peace and conciliation and for a more ready perception, of the harmony which exists between those deductions from our primary intuitions before alluded to, and the teachings of physical science, as far, that is, as concerns the evolution of organic forms—*the genesis of species.*
>
> The aim has been to support the doctrine that these species have been evolved by ordinary *natural laws* (for the most part unknown) controlled by the *subordinate* action of "Natural Selection," and at the same time to remind some that there is and can be absolutely nothing in physical science which forbids them to regard those natural laws as acting with the Divine concurrence and in obedience to a creative fiat originally imposed on the primeval Cosmos, "in the beginning," by its Creator, its Upholder, and its Lord.[11]

In other words, evolution was not the problem; Darwinism was the problem. The problems with Darwinism were not just theoretical. The same year that the *Descent* and Mivart's *On the*

Genesis of Species were published, Darwin's daughter Henrietta
was married to Robert Litchfield. The couple had no children,
but threw themselves into advancing eugenics, which would
become a family cause, replacing their earlier abolitionist passion. Leonard Darwin was named president of the First International Congress of Eugenics, which had been organized by
Francis Galton's Eugenics Education Society. Galton was
Charles Darwin's cousin and was the man who coined the term
"eugenics."

But that was long after Charles's death. While Charles was
alive, his son George Darwin was already showing himself to be
a good eugenicist, championing easy divorce and contraception
to cut down on undesirable traits being passed on. His eugenic
article of 1873 ignited Mivart's ire, and Mivart fired back a
response, accusing George Darwin of promoting licentiousness.
To be fair, George Darwin put forth his suggestions about easy
divorce and contraception purely for eugenic reasons—as if this
were a good excuse—and not to promote sexual promiscuity.
Though Darwin seemed obsessed with sexuality insofar as it
carried forth heredity, he and his family were quite sexually Victorian in all other respects. Charles was deeply insulted on
behalf of his son, and considered a lawsuit on George's behalf.
While the lawsuit didn't transpire, Darwin's friends (Huxley in
the lead) had their revenge on Mivart, nixing him from mem-
bership in the Athenaeum Club, a prestigious scientific society.

In the midst of all this controversy, Darwin turned again to
detailed research, this time in an attempt to link human facial
expressions to man's animal origins. As always, he looked in the
smallest of details for further evidence of his grand evolutionary

conclusions, a movement in the ears akin to a dog's, some connection between laughter and a simian smile, a narrowing of the eyes in angry man and beast. He worked quickly, because he had been taking notes for the subject for years, and published *On the Expression of the Emotions in Man and Animals* near the end of 1872.

All the while, Darwin's fame grew. He became completely identified with his theory; in fact, with evolution itself. In that identity, that fusion of one man's account of evolution with evolution as such, Darwinism was born. The man and his theory became one in the popular mind, and since the popular mind is both the cause and effect of the more sophisticated minds, evolution came more and more to be defined in exactly the way that Darwin demanded. Even if the particular "mechanism" of natural selection was criticized as inadequate to the task (which it soon was[12]), the fundamental assumption that evolution had to occur by means that excluded God became a law of science.

With the growth of fame came the increased bother of letter writing. Darwin, I think, was simply too polite to let anyone's correspondence go unanswered, and he was magnanimous to a fault with his time. He also received interesting gifts from admirers, such as a copy of *Das Kapital* from Karl Marx in 1873. Marx considered him an important ally in the fight against any immaterial residue in human self-understanding, even though as a thoroughly doctrinaire enthusiast of his own theory, he believed that Darwinism should be properly subsumed in the vast historical juggernaut of dialectical materialism. Darwin was either too old or amused to disagree. Surely given his Olympian struggles with German, which he found every bit as frustrating as Mark Twain, *Das Kapital* found its way quickly to the shelf unread. Probably both Darwin and Marx would have been amused by the devout Marxist Joseph Stalin's attempt to cross Darwinism and Marxism in the flesh, creating a new breed, an ape-man soldier and worker, a "living war machine...a new invincible human being, insensitive to pain, resistant and indifferent about the quality of food they eat."[13]

Darwin was busy again with another aspect of eating, not the eating habits of insensitive man-apes, but of sensitive carnivorous plants. His patient research was published in the summer of 1875, *Insectivorous Plants*. Belying his age and frailty, Darwin thought the cessation of work was itself death, so he threw himself into the study of orchids again. Amidst his continual immersion into questions of sexual generation and heredity, his own children had yet to provide him with any grandchildren. This was a trial. Darwin loved his own children, and if ever there was a man made to dote without compromise or scruple upon grandchildren, Darwin was it. Finally, Francis and his wife Amy delivered the news that Charles and Emma would be grandparents, but sadly, death struck with life, and Amy died two days after giving birth in September to Bernard Darwin. Despite the tragedy, or perhaps because of it, Charles Darwin fawned over little Bernard, certain that he was the most extraordinary creature ever to enter the world.

It was during this time that Darwin began writing down casual reminiscences of his life, an incipient self-portrait that was to become his *Autobiography*, a book written primarily for his family, but which came to distort our picture of Darwin for many decades thereafter. It is wonderfully illuminating, but as we've shown, sometimes very misleading.

William Darwin finally married in 1877, and his union, like his sister Henrietta's, remained childless. His brother Horace Darwin was married in 1880. The union was almost blocked by the bride's father who thought Horace too sickly a groom. But Darwin smoothed that over by providing a dowry for the bride. His reward was a second grandchild, yet another Erasmus in the Darwin family tree.

Still, Darwin worked, producing detailed studies on plant variation, the power of movement in plants, and oddly enough, a biography of his grandfather at the very end of the 1870s. Then he turned to the study of worms, publishing his last scientific treatise in 1881. He was sure that they had the rudiments of intelligence as well as sensation, and he went about harassing them to arouse a response, even having Emma serenade them on the piano.

Two months after his brother Erasmus died (in August 1881), Darwin's book on worms came out, and to everyone's amazement, was a big seller. Darwin, however, was sinking. On April 19, 1882, death, the great creative force of evolution, finally came to call on Charles Darwin. Hooker, Huxley, and Wallace were among the pallbearers to his final resting place, in Westminster Abbey next to Sir John Herschel, the famed astronomer who rejected Darwinism, near the eminent Charles Lyell who would only accept a modified form of it, and close to Sir Isaac Newton whom it would have horrified.

Chapter 7

What to Make of It All?

Many have tried to make Darwin a secular saint, whose every fault is politely left unmentioned or hastily excused. Some have tried to make him a demon, and fiercely ignore his virtues. I have tried to paint him as I think he really was.

If our picture of Darwin is still ambiguous, the fault is largely his, rooted in the contradiction of his character. He was an unfailing gentleman, the kindest of husbands, a most loving father, charitable to the poor, sincerely and deeply sympathetic to the sick, the dying, and the bereaved. He bore his own suffering humbly, patiently, and cheerfully where many other men would have crumbled. He envisioned a rosy, progressive future, where suffering would be reduced and virtue would increase indefinitely, where ignorance would give way to knowledge. Finally, he was honest about the strengths and the weaknesses of his theory, and—far more clearly than his disciples—understood its hypothetical character, took criticisms seriously, and treated serious critics with respect. If that were the full Darwin, he might indeed deserve the secular sainthood some would like to award him.

Yet, we have found Darwin to be disingenuous, even with himself. He had an unhealthy passion to be original, so strong that he was willing to ignore, underplay, and dismiss his intellectual predecessors, even the contributions made by his own grandfather. Part of this can be put down to blindness. Darwin lived a kind of insular intellectual existence. His was a close-knit family, and at least all the menfolk took for granted the self-evident truths of Enlightenment skepticism. The skepticism toward Christianity included an evolutionary account directed against the Christian, biblical doctrine of creation. It was part of the comfortable truisms passed on as a heritage. This family heritage allowed Charles to breathe in evolutionary doctrines that had been in the air for over a century—many centuries, if we trace evolution back to the Roman philosopher Lucretius—almost without noticing that the idea had already significantly formed him before he "discovered" it. The idea was so familiar, perhaps, so imbedded in his unconscious assumptions, that when he finally focused his conscious efforts on it, he really may have felt like he discovered it.

But we must not make too many excuses. The facts speak otherwise. Charles Darwin was a third generation evolutionist. He carefully read his grandfather's *Zoönomia* very early on, he studied under the radical evolutionist Robert Grant while in medical school, he worked through the arguments of the French evolutionist Lamarck, and it would be hard to imagine him not discussing evolution with his father and brother around the table and in front of the fire—all this, before he had set foot on the *Beagle*.

We must emphasize this point because it makes a great difference in how we understand his theory (and, I wager, gives us a clue about why Darwin was less than honest in pushing his originality). It means that the theory came before the facts. It was a philosophical and cultural inheritance *before* Charles Darwin himself went in search of evidence to support it. More exactly, it was a *particular* philosophical and cultural inheritance, one that championed the Whig history of rational, secular progress over irrational, religious superstition. Many have pointed out how Charles Darwin's "mechanism" of natural selection fit all too neatly into the more comprehensive social, political, and intellectual context of the late eighteenth and early nineteenth centuries. But that is only part of the story. It fit just as neatly into the even more comprehensive secularizing movements that stretched back to the mid seventeenth century (and beyond). For example, Charles Darwin's account of religion as an irrational reflex of ignorance and fear in the *Descent of Man* was already thoroughly hashed out during the seventeenth and eighteenth centuries, and in fact, was simply a revival of ancient Greek and Roman pagan philosophical views that downgraded all religion to foolish superstition. These views were revived by the Enlightenment, and used directly against Christianity.[1] Darwin's account of the origins of religion was not the *result* of thoroughly sifting anthropological, evolutionary, and historical evidence; rather, the two-century old secular *goal* of eliminating religion by disparagement was the *cause* of people like Darwin searching for evidence to support it.[2] The origin of Darwin's account of religion is not the *Origin of Species*.

But what difference does that make to our assessment of the *Origin of Species*? We have gone to some length to correct another aspect of the Darwin myth, one that he himself helped to perpetuate. The myth places the entirely secular evolutionary approach of Darwin against the irrational approach of scriptural literalists, and asks us to choose: Darwinism or nothing. *Either* a systematically Godless account of evolution *or* a young-earth creationism that sees every warbler and butterfly as being immediately created by God. Species are *either* generated entirely by random variation and natural selection, *or* every last one of them has been immediately and quite recently created miraculously by God. This is precisely what Mivart criticized Darwin for: setting up a false all-or-nothing alternative. Many of Darwin's friends and allies knew the Earth was old, they understood quite clearly the fossil record, they did not try to twist geological evidence to fit the story of Noah's flood, they understood the power of Darwin's theory of natural selection— yet they thought his account of evolution was too small. The principle of natural selection *by itself* ran into too many contradictions; for all its strengths it had too many weak spots. Darwin was never able to explain satisfactorily why so many transitional species were missing from the fossil record; why so many species suddenly appeared in the Cambrian period without any fossil ancestors at all; how small variations, so gradual as to be imperceptible, could be selected as beneficial; how natural selection alone could produce organs as complex as the eye on so many different evolutionary branches, let alone one branch; how radical transformations that occur in the life cycle of one species, such as occur when a caterpillar transforms into a butterfly, could ever have been naturally selected, step by step; why—if environmental conditions have varied so greatly over time and creatures are always varying under the pressure of natural selection—there are "living fossils," creatures that have not changed significantly over hundreds of millions of years, like the crocodile, alligator, cockroaches, dragonflies, and so on; how natural selection could bring about changes in traits that depended on changes in a multitude of other traits to be effective and hence provide a benefit (as, for example, the eye, that depends on a properly curved lens and muscular structure, as well as a vast nerve network hooked into the brain to bring about the experience of vision). Above all, Darwin's account of evolution had to be expanded to account for our peculiar moral and intellectual abilities. These, too, were facts.

This leads to a very important insight that we have brought up before, but is well worth our attention again. Darwin's principle of natural selection was chosen by him precisely *because* it excluded any creative action by God. That is why he was so upset with Lyell and Wallace, and murmured against Gray. They kept letting in God. We should not be fooled by his sop about a Creator added to the *Origin*. As he quite candidly admitted to his friend Hooker, "I have long regretted that I truckled to public opinion & used [a] Pentateuchal term of creation, by which I really meant 'appeared' by some wholly unknown process."[3]

Darwin's systematic exclusion of any divine causation was not called for by the facts, nor is the notion of systematic exclusion of divine causation a necessary presupposition of science as such. Wasn't Henslow a scientist? Sedgwick? Lyell? Wallace? Gray? Mivart? Are we only to allow Huxley the name precisely because he wanted to exclude the possibility of natural theology? Was Huxley really a better anatomist than Cuvier or Owen? The truth of the matter is this: the methodical exclusion of divine causation was an assumption deriving from the particular secular Enlightenment goal of systematically excluding the divine as a matter of human progress. Darwin shared that vision and hence that goal, and it determined the *way* that he defined evolution.

That was the problem with Darwin's theory, and that is the problem with Darwin*ism*. Darwinism is not a synonym for evolution. Darwinism is a particular approach to the evidence for evolution, a reductionist, materialist approach that excludes the Divine on principle. Evolution is a complex and difficult thing we are still trying to understand. Gray, Lyell, Wallace, and Mivart were trying to understand evolution; Darwin was attempting to establish Darwinism, and he was continually frustrated with the defections of his allies.

These considerations have serious ramifications, and not just for how we understand the history of science. Darwin's whole racial and eugenic vision in the *Descent* was the direct result of his insistence that God be excluded from his account of evolution. There were other ways to be a racist, and Christians, to their shame, played the game by linking "lesser" races to outcast figures in the Bible, or like the naturalist Louis Agassiz, Harvard's leading paleontologist, by contriving different creations for different human races. But Darwin, incensed that his own allies would insist that natural selection could not explain the evolutionary development of human morality and intellectual capacities, turned his formidable intellectual powers to an entirely reductionist account of humanity.

The only way to make everything depend on natural selection was to argue that our most human traits were built up, bit by bit, in the struggle of tribe against tribe, race against race, where the losers were extinguished and the winners advanced another rung upwards. This led immediately to a ranking of the existing races, since they must fit the evolutionary model and fill in the hierarchy of ape developing into European man. In order to make his case, Darwin had to fashion the "lower" races as more apelike and the higher apes as more human so as to close the evolutionary gap. And we must remember that in making moral capacities the direct result of natural selection, Darwin had then to argue that the moral capacities of each existing race were, like any evolved traits, determined within strict limits. The "lower" races—the Negro, the Australian, the Fuegian—were morally blunted, morally inferior. As *existing* races they could not be expected to be morally improved for the same reason that a particular Beagle cannot be trained to outrace a greyhound. It might be possible in the future, through breeding many successive generations, to develop a faster dog from a slower Beagle, but the original Beagle has already had his cards dealt to him. In the same way, it might be possible to improve the breeding stock of the savage sometime in the future, but improvement only belongs to *future* generations. The "lower" races existing in the present are as good morally as they are going to get. Natural selection has already dealt them their hand. The same is true of their intellectual abilities. In both traits, moral and intellectual, natural selection has only given *already existing* individuals and species a certain level of attainment, and no more.

To put it another way, for Darwin apes do not evolve into men; rather, *some* apes, way back when, were (for example) a little more intellectually endowed by natural selection than other, similar apes, and the greater exterminated the lesser. By a process of minute increase and ruthless extermination, the evolutionary tide rises slowly from super-ape to sub-human, minute step by minute step, the slightly more fit eliminating the slightly less fit. There are no leaps for Darwin, so existing "savages" have hit their evolutionary high point, both morally and intellectually. Improvement is a possibility only for future generations.

In saying all this, Darwin wasn't blinded by being a "man of his time." He should, indeed he could, have known better. Didn't Darwin see the improvement among the Tahitians that Christianity—and not natural selection—brought about? Perhaps he was swayed against this evidence by the experience of having the trappings of civilization fall so easily from the Fuegians, Jemmy Button, Fuegia Basket, and York Minster. Or, more likely, he simply ranked the Tahitians as a more advanced race than the Fuegians.

Be that as it may, certainly his theory's co-discoverer, Alfred Wallace, gave him hard enough evidence that natural selection was insufficient to explain the arrival of humanity. Wallace, who had a lot more direct experience of "savages," argued that these seemingly "undeveloped" human beings actually had the same, or nearly the same intellectual powers as a European—and this despite the fact that they lived at a level little above the animals. For Wallace, this was sterling proof that all human beings, however uncivilized they may appear, shared in certain human capacities that clearly set them far above the apes. That "gap"—between ape and man—was too large a leap for natural selection to explain, and the "savages" couldn't be used to fill it. In Darwin's view, if savages lived at a level only a hair above the apes that was a sure indication of the watermark to which evolution had lifted them. But, reasoned Wallace, if savages had intellectual capacities far beyond what their environment would demand, then something greater must be at work. Darwin would have none of it, and that obstinacy brought him to rank the existing races, making of them an ascending ladder to bridge the gap.

But the cost of this obstinacy was extraordinarily high, so high that Darwin lied to himself about the implications. As his friends warned, as Wilberforce warned, as his opponents warned, the cost of denying that *fact* of peculiar human moral abilities is that morality itself recedes into the shifting sands of aimless variation and ruthless selection. Despite Darwin's optimism in regard to evolutionary moral progress, the arguments of the *Descent* only illustrate, in the most alarming way, that the notion of "survival of the fittest" is a brutal tautology.

What do I mean by that? Many have commented on the philosophical difficulty entailed in the notion of the "survival of the fittest." It is a tautology; that is, it looks like it is explaining something when in reality it suffers from a fatal redundancy and explains nothing. It defines the "fitness" of something, not in any positive and informative way, but simply by the fact that it is not dead. Not-deadness doesn't go very far as a biologically illuminating principle. But when we apply it to the evolution by natural selection of moral qualities, the tautology becomes brutal. Not-deadness now means moral superiority because moral superiority has been reduced to whatever habits or social customs contribute to the survival of a particular individual, tribe, race, or nation in the ongoing struggle against other individuals, tribes, races, and nations: in short, might makes right. We return again to the issue of slavery, because Darwin rightly took it to be *in fact* a moral evil. It has been argued that Darwin's affirmation of common ancestry for human beings was formed in great part by his hatred of slavery.[4] On this view, Darwin was willing to bend science to a worthy moral cause: common ancestry proved that white man and the black man shared a common origin, and so were brothers of sorts. Perhaps this is right about Darwin's intentions, but only goes to prove our point when we look at the results. Common ancestry doesn't keep slavery from being natural. Natural, in Darwin's evolutionary scheme, means according to the principle of natural selection. There is no doubt that all ants, slaving and non-slaving, have a common ancestor, and that natural selection produced both variant species—not by taking a wrong turn and a right turn, but simply by branching off. As Darwin very carefully argues, animal instincts and human moral habits and social customs exist on one continuum. Human moral habits and social customs differ according to the principle of natural selection. There is no doubt that all men in all human societies, slaving and non-slaving, have a common ancestor, *and* that natural selection has produced these social variants, not by taking a wrong turn and a right turn, but simply by branching off. There *is* no wrong or right turn in evolution, so there can be none in human evolution as Darwin understood it. Whatever contributes to a society's self-preservation is affirmed by natural selection. As Darwin was so fond of saying, natural selection is blind, and that is not the blindness of the blindfold of Lady Justice, but the blindness of a natural process that begins with random variations and ends with the declaration of the winner by the death of the loser. If the slavery of one race by another contributes to its victory in the struggle to survive, then natural selection has heartily and heartlessly affirmed slavery as natural and effective. Nothing can condemn what works. We repeat what Darwin himself insisted upon: if there were a moral standard outside the process of natural selection, if the evolution of morality progressed toward that standard, if the actions of men and societies were judged by that standard, then we would be admitting a theistic account of evolution. And that would mean random variation and natural selection would be insufficient to explain evolution.

Darwin would not allow that, but he also could not face the completely amoral consequences of his own theory. He tried to slip in a goal without God, proclaiming that evolution was culminating in the moral trait of "sympathy." What Darwin meant—or better, all he *could* mean—was that evolution was leading up to Darwin himself, to a man with a very soft heart, a good family man, kind husband and father, a man who couldn't abide cruelty to human beings or animals, a good English gentleman, a humanitarian liberal Whig, a stalwart opponent of slavery. But what does *that* mean? Human evolutionary history becomes autobiography.

But that will not do, or better, it cannot fit into Darwin's own account of evolution. In evolution, wherever you are, that's where you'll be. Of course human evolution led up to Darwin and *his* morality, but in precisely the same way that it led up to everyone else who was in the historically privileged position of being not-dead and to his or her respective morality as well. The science of ethics can only be genealogical research that follows out the separate branches; no branch is privileged. And since "natural selection is daily and hourly scrutinising, throughout the world, the slightest variations; rejecting those that are bad, preserving and adding up all that are good; silently and insensibly working, *whenever and wherever opportunity offers*, at the improvement of each organic being in relation to its organic and inorganic conditions of life,"[5] the branching will continue. A branch does lead up to Darwin, and whatever fine moral qualities he could boast, but then it leads right past him as well—not up, but wherever. As with any evolved trait, we cannot judge the different evolved moralities by any other criterion than survival: that's all "good" can mean. A lack of sympathy, a kind of coldness in using hard reason, an unblinking adherence to the dictates of ruthless nature, a eugenic imposition of the principle of the survival of the fittest to the unfit, might all prove beneficial in future struggles, especially against soft, sympathetic, and sickly Englishmen. The outcome of the various battles among the branches will determine how the human evolutionary tree is pruned, and hence the trajectory of the next branches in turn. But by Darwin's own principles, Darwin's own "morality" is no more moral than the shape of his nose or the color of his skin.

So we must keep clear the distinction between *wanting* something to be evil and having an actual firm foundation in reality for something actually *being* evil. Darwin's worst lie, the one he told himself, was based in a kind of self-deception that he could have his moral cake and eat it too. In the *Descent*, he allowed all foundations for morality to be consumed by natural selection, and then he tried to sneak "sympathy" back in as the moral trait at the apex of evolution. At best, that was dishonest. At worst, it allowed him to rank all other societies below his own as ripe for evolutionary elimination.

What are we to make of it all, then? Not that Darwin was a Nazi. There is something both perverse and tiresome about our notion that Nazism is the only real evil, and consequently that the only way to demonstrate the evil of anything is to trace it to or from the Third Reich. The fundamental problem with Darwinism is not that it leads to Nazism, but that it can lead to anything.

One branch did in fact lead to Nazism, and Darwin provided the Nazis with a eugenic framework for the extermination of the biologically inferior, and for the extermination of Jews and Gypsies, the "extinction" of less morally and intellectually fit races at the hands of the more fit.[6] (We shall fill out this connection in the next chapter, the Postscript.) Again, Darwin would have been horrified, but Darwin personally *not wanting* a particular evil must be distinguished from the *framework* of his theory, which not only allows eugenics and genocide, but regards them both as implicit to the workings of natural selection.

But the branch leading to Hitler was just one possibility. The branches can lead anywhere and everywhere because the only "moral" penalty is dealt out by natural selection itself: self-destruction. Individual or societal self-destruction isn't evil. It's nothing at all—just a dead-end branch. The transgressors aren't around anymore to receive their penalty. Yet we must add that human beings themselves are agents to natural selection, as Darwin's account of the rise of our intellectual and moral capacities through the conquering and extinction of tribe by tribe, race by race makes clear. If one society crushes another, that is not wrong. That is not even a shame. That *is* natural selection at work.

We repeat, on Darwin's own terms, human evolution can have no moral goal, and there is no sure way of predicting where the various branches might go. A society could very well flourish by oppressing other weaker societies; by systematically eliminating the retarded, the deformed, those genetically disposed to certain diseases in the womb, and the elderly and sick in their beds; by having a government-supported program of contraceptive education and dispersion in poor neighborhoods to cut down on racial minorities; by cultivating a vast network of cyber-pornography for financial gain; by mesmerizing the population with mind-numbing titillation so it is more easily controlled; by engaging in high-tech cannibalizing of medical, medicinal, and cosmetic tissue from human beings artificially conceived and grown for such purposes. If such a society could exist, then by the very fact that it exists, we must say that it has survived, and hence that it is good—even when our better natures, so long as we can remember them, tell us it is not.

We have examined Darwin's life, and his lies. It is unpleasant to bring the charge of lying against so revered and gentlemanly a figure, but it must be done precisely because Darwin's influence over our own lives is so great. I have mentioned that, in my opinion, the worst lie was the one he told himself, that he could have his moral cake and eat it too, pushing forward a godless account of evolution that made morality a mere transient effect of natural selection, and at the same time holding up particular moral traits, such as sympathy, as if they were somehow the aim of aimless evolution. Perhaps I was somewhat hasty in ranking this as the worst lie, because it really is the result of another, that evolution *must* be understood as godless, that is, that evolution must be Darwinian evolution. That insistence was not, itself, scientific; it was a philosophical prejudice that defined science in obstinately secular terms; a prejudice, I have argued, that comes from particular Enlightenment assumptions. It was the same Enlightenment prejudice, so comfortably ensconced in the Darwin family line, that brought Darwin to tell another lie in his *Autobiography*: that he was, as a young man, a scriptural fundamentalist, a stout believer in Anglicanism's Thirty-nine articles, and only lost this faith as a result of the evidence seen while sailing the world in the *Beagle*. This, I maintain, was simply Darwin fitting his life into the standard Enlightenment Whig history of progress from superstition to science, one that still almost entirely forms our own imagination and hence historiography. It is not a small lie, because it contributes to the notion that evolution must be godless or it cannot be scientific. What Darwin's life proves, however, is that if evolution is godless then it cannot be moral. It proves it, not because Darwin *himself* was immoral, but because Darwinism cannot help but collapse morality into the survival of the fittest.

Perhaps the most important distinction we have recovered in our account of Darwin is this: the distinction between evolution and Darwinism. As we noted, Darwin tended to put his argument in either-or terms: either you accept Darwinism or you must embrace fundamentalism. But that these are the only two alternatives is, well, a lie. His own friends thought otherwise, and insisted that, while Darwin's arguments in regard to natural selection were ingenious and explained a good deal, natural selection was insufficient to produce the actual complexity of nature we see before us, especially the moral and intellectual complexity of human beings. After all, if *we* don't count as evidence, what does? But this position—embraced, in its several ways, by Lyell, Wallace, Gray, and Mivart—is not taken seriously today. This has nothing to do with the actual evidence, but everything to do with the Enlightenment Whig version of history that colors how we view the world; many of us cannot help but assume that Darwinism displaced Christianity in explaining our origins, just as surely as science is always triumphing over superstition; opposition is simply dismissed as ignorance—as in the great mythmaking play *Inherit the Wind* about the Scopes "Monkey Trial"—even if, as with Lyell, Wallace, Gray, and Mivart, ignorance is manifestly not the case.

We are still caught in the lie that Darwinism is the only respectable scientific position. I wish I could say that Darwin himself was not that small-minded, but I am afraid that he was, which is a painful thing to discover about so gifted a scientist, so kind and compassionate a father and husband, so devout an abolitionist, so amiable a friend as Charles Darwin. I shall pay him the greatest respect by saying that I am convinced beyond any doubt, that he was greater than his theory, both morally and intellectually, and that the only theory of evolution worth holding is one much grander, more intricate and elaborate, one far more subtle, complex, and manifold; in short, one into which Darwin the man would fit.[7]

Chapter 8

Darwin and Hitler

The distinction between Darwin the man and Darwinism the particular evolutionary theory is all-important. While Darwin the man would have been horrified at Hitler and the Nazis, Darwinism proved all too influential in providing a scientific foundation for their racial and eugenic theories.

Darwinism, of course, was not the *sole* cause of Nazism. That would be absurd and inaccurate; among the many tangled roots of Nazism, one would have to mention the messianic nationalism that became deeply imbued in Germany during the Reformation; the mystical-mythical notion of Germany's pre-Christian pagan roots that exalted the noble Teutonic warrior barbarian; the cultural and political anti-Semitism that had infected Germany at least since the time of Luther; the long-standing militarism of German society; and the humiliation Germans felt after the First World War and in the midst of the Great

Depression. But all that being said, Darwinism was a significant contributing factor.

In Darwinism, German intellectuals found scientific vindication that racial conflict, or more exactly, the subordination or elimination of inferior races, was the one needful task to save the world from evolutionary degradation, and even more, to advance humanity physically, morally, and intellectually. These were not ideas that German intellectuals twisted out of context from ill-conceived offshoots or aberrations. They came straight from Darwin himself.

Darwin argued that human beings evolved through the struggles of races and tribes, with the most fit exterminating the least fit—a process that continued and implied the eventual disappearance of such as the aboriginal Australians. Darwin believed that man was "the very summit of the organic scale" not because he been "aboriginally placed there" by God, but because he had triumphed through natural selection, a fact which "may give him hopes for a still higher destiny in the distant future."[1]

Such is the thrilling finale of Darwin's *Descent of Man*. But if human evolution is to advance, it can do so only through the same evolutionary means, through the extermination of "less favoured races" by those that are more "favoured." If human evolution can advance, it can also retreat or recede, slipping backwards if the number of less fit is allowed to increase at the expense of the more fit. In Darwin's summary statement, "Man, like every other *animal*, has no doubt advanced to his present high condition through the struggle for existence consequent on his rapid multiplication; and if he is to advance still higher he must remain subject to a severe struggle. Otherwise he would soon sink into indolence, and the more highly-gifted men would not be more successful in the battle of life than the less gifted."[2] We must remember, Darwin warns us, that "progress is no invariable rule," and if we do not "prevent the reckless, the vicious and otherwise inferior members of society from increasing at a quicker rate than the better class of men, the nation will retrograde."[3]

Thus spoke Darwin himself, and as his arguments spread all over Europe and America, so did an obsession with notions of racial gradation and racial degradation. It was *not*, we must stress, solely a German phenomenon; in fact, it was very much an English and American one as well. The very word "eugenics" was coined by Darwin's own cousin, Sir Francis Galton, and Darwin's family took a lead in the eugenics movement. So did America's Margaret Sanger, foundress of Planned Parenthood, who spearheaded an international birth control movement to keep the "inferior members of society from increasing at a quicker rate than the better class of men."[4] The subtitle of Darwin's most famous book, *The Origin of Species*, is *The Preservation of Favoured Races in the Struggle for Life*, and racism and eugenics have always gone hand in hand.

While Darwin did not call for exterminating Jews, he did set forth a theory in which racial extermination is the engine of evolutionary progress. That is how he could, with all equanimity, speak of future racial extermination as entirely predictable. We return again to his shocking words, words worth repeating because they are so often glossed over or ignored by other biographers. According to Darwin, "Extinction follows chiefly from

the competition of tribe with tribe, and race with race," pushing man up the evolutionary ladder. "When civilised nations
come into contact with barbarians the struggle is short, except
where a deadly climate gives its aid to the native race,"[5] Darwin
wrote in his *Descent of Man*, a little more than a decade after
publishing the *Origin of Species*. "At some future period, not
very distant as measured by centuries, the civilised races of man
will almost certainly exterminate and replace throughout the
world the savage races. At the same time the anthropomorphous
apes will no doubt be exterminated. The break [between human
beings and apes] will then be rendered wider, for it will intervene
between man in a more civilised state, *as we may hope*, than the
Caucasian, and some ape as low as a baboon, instead of as at
present between the negro or Australian and the gorilla."[6]

There you have it. Not only is racial extermination central to
human evolution as it has gotten us to where we are today, but
it remains central in the exciting task of climbing "as we may
hope" higher "than the Caucasian," leaving far behind those
evolutionary racial steps, the "negro" and the "Australian"
along with the gorilla. So goes the ascent to the evolutionary
übermenschen. Darwin was partial to considering Englishmen
rather than Germans as the chief candidates for übermenschen-
hood, but that was a matter of national pride rather than science, and it was the purported science that was important,
because Darwinism provided powerful, *scientific grounds* for
anti-Semitism.

We must also remember that there was more to the Nazi program than the extermination of the Jews. Also marked for elimination were Slavs and Gypsies and the handicapped or "unfit."

It is unfair to say that Darwin would have approved of such "hard" eugenics, since we know that he personally believed that sympathy for the human "unfit" must win out over "hard reason." Yet, as we noted, holding up sympathy as a kind of moral goal is undermined by Darwin's insistence that evolution can have no goal, and that all evolved traits—sympathy included—are as transient as the conditions that make them useful. If "the reckless, the vicious and otherwise inferior members of society" begin to outnumber the "better class of men," so that "the nation" suffers evolutionary "retrograde" rather than progress, it would seem that hard reason should replace soft sentiment. Whether Darwin himself would personally affirm it is of no consequence since the principles of Darwinism both allow and encourage morality to be based upon natural selection, where the stronger rise through the extermination of the weaker.

It is certain that in Germany, the struggle between sympathy and hard reason, however protracted, was won by hard reason, hard *scientific* reason. That is, it is *not* the case, as it is too often assumed and argued, that something called social Darwinism broke off from biological Darwinism, and as a pseudo-scientific movement crowded with know-nothing thugs illicitly seized upon Darwinian biological concepts and illogically applied them to the social realm in Germany. There is no real distinction between Darwinism and social Darwinism, except in the merely accidental distinction of being grounded in two separate books, Darwin's *Origin of Species* and Darwin's *Descent of Man*. More accurately, as historians have pointed out, social Darwinism predates Darwin's signal literary contributions, and is traceable to Thomas Malthus, Herbert Spencer (who actually gets the honor of coining the term "survival of the fittest"), and Darwin's cousin Francis Galton. But that admirably illustrates exactly my point in previous chapters, that Darwinian evolution is not Darwin's discovery, but the culmination of (at least) a century-long intellectual movement that was attempting a top-to-bottom reformulation of human biological and social life. Darwin put his stamp on it, and it bears his name, so he too must bear the responsibility for the way he himself applied his particular version of a reductionist evolutionary theory to human beings.

Having said all that, there is still the nagging question, how did Darwinism come to such a brutal realization in the policies of Nazi Germany? As we've noted, Darwin made sure that his *Origin of Species* was translated into German as soon as possible. One of its earliest and most influential readers was Ernst Haeckel, who had earned doctorates in both medicine and zoology. After reading Darwin, he was instantly transformed into a devoted disciple, even trekking off to Down on pilgrimage (much to the chagrin of Emma Darwin, who could not take his bellowing in bad English). Darwin said to Haeckel—most significantly—that he considered him, of those who were carrying on his work, "to rank as the first."[7]

Haeckel, the eminent zoologist, was an indefatigable disciple. The scientific popularizations he provided the German public sold in the hundreds of thousands, and were translated into twenty-five languages. Haeckel united his Darwinism to the notion of a magic, pre-Christian, Germanic pagan past. He believed the science of Darwinism combined with the virility of Germanic paganism could root out the contagion of Christianity, and produce a new and masterful Germany. Liberating Germany from Christianity was a hobby horse of much of Germany's (and Europe's) intellectual class, and Germany had an especially well-educated population, which was open to such intellectual currents. This liberation, of course, meant liberating Germany from Christianity's moral confines, specifically shedding Christian sympathy in favor of evolutionary hard reason.

Haeckel, on this point, offered an implicit criticism of Darwin himself: that Darwin's notion of sympathy being the result of evolution was wrong; rather, sympathy was the result of Christianization. Christianity's unscientific belief that each human being has a human soul—the origin of its notions of charity and sympathy for the "unfit"—could not hold up against the new scientific materialism. It *must* not hold up because such "charity" causes racial degradation. As it is "practiced in our civilized states" it results in the "sad fact that . . . weakness of the body and character are on the perpetual increase among civilized nations. . . . " "What good does it do to humanity to maintain artificially and rear the thousands of cripples, deaf-mutes, idiots, etc., who are born every year with an hereditary burden of incurable disease?" Haeckel thundered. "It is no use to reply that Christianity forbids" their destruction because Christianity's opposition "is only due to sentiment and the power of conventional morality" and "pious morality of this sort is often really the deepest immorality." Against Darwin, but not against Darwinism, Haeckel was quite clear that "Sentiment should never be allowed to usurp the place of reason in these weighty ethical questions."[8]

But it would be a mistake to think that one man was responsible for the application of hard Darwinian reason to the social realm in Germany. Darwin's thought quickly penetrated the highest levels of the German intellectual realm, and was espoused by its most eminent academics. Robby Kossmann, a zoologist and later medical professor, stated quite candidly in 1880 that "the Darwinian world view must look upon the present sentimental conception of the value of the life of a human individual as an overestimate completely hindering the progress of humanity. "The human state, as an embodiment of race, an organic unity, "must reach an even higher level of perfection . . . through the destruction of the less well-endowed individual, for the more excellently endowed to win space for the expansion of its progeny."[9] As Alexander Tille, a disciple of Haeckel and a prominent Darwinian in his own right, stated very succinctly, the new moral goal must be "the elevation and more excellent formation of the human race," and this means that "even the most careful selection of the best can accomplish nothing, if it is not linked with a merciless elimination of the worst."[10] He was also a devotee of the highly influential anti-Christian German philosopher, Friedrich Nietzsche who asserted that the "new party of life," which has as its "greatest of all tasks" the "higher breeding of humanity," must pursue "the unsparing destruction of all degenerates and parasites."[11]

At the dawn of the twentieth century, Friedrich Krupp, the prominent industrialist of the Krupp firm that armed Germany, announced the lucrative Krupp Prize would go to the best answer to the question, "What do we learn from the principles of biological evolution in regard to domestic political developments and legislation of states?" The expected answer—loaded as the question was, by the adamantly eugenic Darwinian enthu-

siast Krupp—was happily given by a physician, Wilhelm Schall-
mayer, who toted home ten thousand marks for his *Heredity
and Selection*.[12]

Alfred Ploetz, who was entirely enamored of both Darwin
and Haeckel, and who was not an anti-Semite, founded the
world's first eugenics organization in 1905, the German Society
for Race Hygiene. Other prominent German Darwinian eugeni-
cists, social Darwinists, and racial theorists were the ethnologist
Friedrich Hellwald; the psychiatrists Hans Kurella, Robert Som-
mer, and Eugen Bleuler; professor of anthropology at the University of Berlin, Felix von Luschan; the physiologist Wilhelm
Preyer; the physicians Ludwig Büchner, Eduard David, and Ludwig Woltmann; evolutionary biologist August Weismann; the
anatomist and anthropologist Eugen Fischer; the zoologists
Heinrich Ziegler and Oscar Schmidt; the geographer Alfred
Kirchhoff; the botanist Ernst Krause; the sociologists Ludwig
Gumplowicz, Sebald Steinmetz, and Klaus Wagner; and the list
goes on. Again, these were not brown shirt thugs, but denizens
of the highest ranks in academia and German intellectual culture, and their work was done largely in the latter half of the
nineteenth century and the first quarter of the twentieth; that is,
they were part of the overall Darwinian worldview that so
formed Germany as a preparative to Hitler and the rise of the
Nazis. It was often anti-Semitic, but it was always Darwinian,
and anti-Semitism was firmly embedded in a Darwinian racial
framework long before Hitler rose to power between the two
World Wars.

In fact, social Darwinism was a major factor in German
thinking in World War I. As one military historian has noted,

citing the biologist (and Darwinian) Stephen Jay Gould, "It was a commonplace among German intellectuals—and the German General Staff, who spoke these sentiments before the Kaiser—that the War was justified by," in Gould's words, "an evolutionary rationale...a particularly crude form of natural selection, defined as inexorable, bloody battle."[13]

Of course, after the First World War, the German National Socialist Workers' Party, the Nazis, took this thinking even farther. It should be illuminating that Rudolph Hess, the deputy party leader of the Nazis, stated matter-of-factly that "National Socialism is nothing but applied biology."[14] Hitler was a social reformer who wanted to cure what ailed society by Darwinian means, combining (in his own words in *Mein Kampf*) "a ruthless determination to prune away all excrescences [in society] which are incapable of being improved"[15] with a positive program of breeding the more fit Aryan race for world domination. The "highest aim of human existence" argued Hitler "is the conservation of race."[16] It is impossible to miss the Darwinian structure and tone of the following words from *Mein Kampf*, where the *völkisch* philosophy is simply a restatement of Darwinian racial ranking, struggle, and survival of the fittest race, and the nation becomes the bearer of racial, evolutionary advancement:

> [T]he *völkisch* concept of the world recognizes that the primordial racial elements are of the greatest significance for mankind. In principle, the State is looked upon only as a means to an end and this end is the conservation of the racial characteristics of mankind. Therefore on the *völkisch* principle we cannot admit that one race is equal to another. By recognizing that they are different, the *völkisch* concept separates

mankind into races of superior and inferior quality. On the basis of this recognition it feels bound, in conformity with the eternal Will that dominates the universe, to postulate the victory of the better and stronger and the subordination of the inferior and weaker. And so it pays homage to the truth that the principle underlying all Nature's operations is the aristocratic principle and it believes that this law holds good even down to the last individual organism.... The *völkisch* belief holds that humanity must have its ideals, because ideals are a necessary condition of human existence itself. But, on the other hand, it denies that an ethical ideal has the right to prevail if it endangers the existence of a race that is the standard-bearer of a higher ethical ideal. For in a world which would be composed of mongrels and negroids all ideals of human beauty and nobility and all hopes of an idealized future for our humanity would be lost for ever.

On this planet of ours human culture and civilization are indissolubly bound up with the presence of the Aryan. If he should be exterminated or subjugated, then the dark shroud of a new barbarian era would enfold the earth....

Hence the folk concept of the world is in profound accord with Nature's will; because it restores the free play of the forces which will lead the race through stages of sustained reciprocal education towards a higher type, until finally the best portion of mankind will possess the earth and will be free to work in every domain all over the world and even reach spheres that lie outside the earth.[17]

Nothing in this entire tirade presents an idea peculiar to Hitler. On the contrary, Hitler was merely drawing together longstanding elements of German Darwinian thought that had been brewing for well over half a century. As he said in a speech of 1923, "All of nature is a constant struggle between power and weakness, a constant triumph of the strong over the weak."[18] Certainly he did it in an arresting way—in part by draping the brutality in religious terms, identifying the brutal laws of natural selection with the laws of God—but the racial and eugenic core of his thought was hardly novel. Nazi barbarism was scientific barbarism. In Richard Weikart's apt words,

> Nazi barbarism was motivated by an ethic that prided itself on being scientific. The evolutionary process became the arbiter of all morality. Whatever promoted the evolutionary progress of humanity was deemed good, and whatever hindered biological improvement was considered morally bad. Multitudes must perish in this Malthusian struggle anyway, they reasoned, so why not improve humanity by speeding up the destruction of the disabled and the inferior races? According to this logic, the extermination of individuals and races deemed inferior and "unfit" was not only morally justified, but indeed, morally praiseworthy. Thus Hitler—and many other Germans—perpetrated one of the most evil programs the world has ever witnessed under the delusion that Darwinism could help us discover how to make the world better.[19]

Again, I have no doubt that if somehow Darwin could have lived to see what became of Darwinism, he would have been absolutely mortified. But would Darwin have been sufficiently shocked to question Darwinism itself? That, I cannot answer.

Chapter 9

Christianity and Evolution

I have labored to make clear one very important distinction, that Darwinism is different from evolution. Darwinism is the name that we properly give the approach to evolution championed by Darwin. Evolution is the thing that happened. Darwinism is a particular theory that defines itself in an entirely reductionist, materialist way to avoid at all costs letting a divine foot in nature's door. Evolution is a fact, the marvelous and still largely mysterious complex of evidence that gives every indication that nature is a spectacular work in progress. This distinction allows me to say a most astounding thing: one can heartily accept evolution on scientific grounds and roundly reject Darwinism on scientific, philosophical, moral, and theological grounds.

This distinction is important for many reasons, but I'd like to focus on one in particular, namely the relationship of Christianity to evolution. Christians today are, generally speaking, divided into three camps: those who reject evolution because they believe that it leads directly to atheism; those who accept evolution and deny that it leads to atheism; and those who are indifferent to the question and go on about their business.

To the last group, I have little to say. Minding one's own business may be, for some, their proper vocation. For others, it is a form of sloth, and that is a deadly sin, the lethal pose of "whatever" to life's most important questions. Be that as it may, I would like to invest more time in a visit to the other two camps.

Those Christians who reject evolution because they believe that it leads to atheism are indeed proceeding from a proper fear. Insofar as Darwinism has swallowed up all of evolution into itself, the evolutionary theory partakes of the deep anti-theistic bias that Darwin built into it. It in fact *does* lead to atheism because it was designed to do so. The enormous push that secularization received from Darwinism should be proof enough that the theory of evolution so understood destroys belief in God. The problem with this camp—if we recall our distinction between Darwinism and evolution—is that its denizens feel they must then attack evolution *itself*, that is, all the evidence from the great age of the earth to the fossils, that indicates all too clearly that God did not create the earth and all its creatures, fully-formed, just six thousand years ago. Needless to say, Christians of this camp appear entirely irrational and unscientific.

But there are plenty of Christians, who take themselves to be of the more sophisticated sort, who accept evolution and deny that it leads to atheism. They blithely ignore the obvious historical fact that Darwinism has been the most significant contributing cause in the de-Christianization of the west, and what should be the obvious contemporary fact, that most evolutionary biologists today (or at least most of the famous and influential ones) are atheists *because* they regard evolution as having proven that the whole God thing is intellectually obsolete. The problem with this camp—if we again recall our distinction between Darwinism and evolution—is that its denizens feel that they must uncritically defend Darwinism *itself*, as if all the evidence must be sifted through an entirely reductionist, materialist filter, and also that they must attack anyone who has any reservations at all about uncritically accepting Darwinism.

The unpleasant result in regard to Christianity and evolution is that confusion reigns in the present debates about their proper relationship, and even moments of clarity that occasionally slip through are howled into eclipse by these two most vociferous camps, each possessed by at best a half-truth.

I would like to introduce someone into this debate, a stranger to be sure. We might call him the reasonable Christian. He is to be distinguished from the Christian fideist who wrongly attacks evolution because he rightly sees the damage caused to the faith by Darwinism, and from the rationalist Christian who wrongly defends Darwinism at all costs because he rightly sees the damage caused to reason by the attack on evolution on behalf of Christian faith. Paradoxically, the Christian fideist, I shall argue, has a stunted view of faith, and the rationalist Christian, a stunted view of reason. Each is actually plagued by the disease that he sees in his opponent.

What about the reasonable Christian? The reasonable Christian holds, first of all, that science cannot contradict the faith because he assumes that the Creator God and the Redeemer

God are one and the same God. He differs markedly in this from both the Christian fideist and the rationalist Christian. The fideist is often driven to deny science that seems to contradict the faith; the rationalist to deny every aspect of faith that seems to contradict science.

The reasonable Christian does not allow that a contradiction is possible on either side. He knows from the history of science itself that science, including evolutionary science, is a merely human activity, and that despite its pretensions, scientists are often wandering in confusion, hobbled by bad theories, and mis-led by their very victories into assuming that they are omnis-cient. He knows that nature, as a creation of the profound wisdom of God, is much more magnificent and mysterious than our human attempts to grasp it, and so assumes that evolution must be something far grander than Darwin made it out to be, something so marvelous that, if we fully understood it, it would appear miraculous—a manifestation of the glory and wisdom of the Creator. Darwinism is too small for him as a theory of evo-lution because nature is too big for Darwinism to be true. A proper theory of evolution would not reduce the real complex-ity of human beings to make them fit within a tight materialist and reductionist framework; it would *expand* the theory of evo-lution so that the real moral, aesthetic, and intellectual com-plexity of human beings, as the pinnacle of evolution, defines the framework to understand all of evolution.

On the other hand, the reasonable Christian thinks even more highly of God's revelation in Scripture than he does of nature, for nature, however grand, is still only an effect of God, whereas Scripture is a revelation of God Himself. The history of how

Christians have attempted to interpret Scripture is warning enough to him that interpretation is no easy thing, and that we must be ever wary of substituting merely human wisdom and ingenuity for the fathomless depths of divine wisdom and genius in the Bible. A strictly literal reading of the Bible is too small for him for the same reason, analogously speaking, that a strictly literal reading of Shakespeare or Plato would be too small—it can't take into account the complex, multi-layered reality of the text. If some genuinely, well-established fact seems to contradict our reading of Scripture, then the reasonable Christian knows that there must be some problem with our reading of Scripture. It is saying *more* than we thought it was, something far more profound and exhilarating. Greater depths are being revealed.

I would be misleading the reader if he thought this all led to an entirely open-ended, and hence undisciplined approach to both evolution and the Bible. There are certain givens that cannot be given up or both science and faith will be destroyed. These are, most importantly, the very givens that Darwin himself gave up so that he could establish an entirely God-less account of evolution.

Our peculiar moral and intellectual capacities are not up for negotiation; they are givens that any account of evolution must explain, not (like Darwin) explain away. The cost of explaining away our moral capacities as mere after-effects of natural selection should now be clear: good and evil are reduced to the brutal level of what contributes to survival and what inhibits it. To live by any means becomes the golden rule. In our own time, that manifests itself in our attempts to continually manipulate human nature as if we were simultaneously both clay and potter.

This notion that we can endlessly tinker with our own nature does not result in making out of ourselves something far nobler. The predictable result is that science simply becomes the slave of our every physical whim and pleasure, and even more degraded, of the desperate and repugnant attempt to produce physical immortality, thereby creating doddering fools frozen in a state of perpetual youth.

The cost of explaining away our intellectual capacities as mere aftereffects of natural selection may be less clear. Insofar as they are a special mark of our humanity, insofar as that mark is erased, we will be increasingly reduced to the moral level of other animals. That may mean that we start treating other animals as our moral equals, granting them rights and privileges that only human beings previously enjoyed. Or it may mean that we start treating other human beings as animals, that is, in entire disregard of their having any special moral nature. When that happens, we may treat them as we would any other animal: shoot them when they are bothersome, and even eat them if we are so inclined, use them for experimentation, and have them as beasts of burden. If you think that is a fantasy, then you don't understand that the reason why, for enthusiasts of euthanasia, taking your dog to the vet to be put down is *just like* having your handicapped baby or ailing grandmother put down in the hospital. If you think cannibalism too distant a possibility, then you do not understand the dark spirit behind embryonic stem cell research. But laying all this aside, it should be obvious that a view of evolution that cannot explain why one particular creature has intellectual capacities that so extraordinarily pass those found in other animals that it can set out a theory of evolution and attempt to prove it through the intense study of nature, using hopelessly intricate scientific instruments that allow it to peer into the mind-numbingly complex microscopic world of the cell—that view of evolution, I say, is so stunted as to be laughable.

Finally, and following directly on these last points, no theory of evolution that assumes on principle that God does not exist can be a valid theory of evolution. By now, I hope, the reader understands the reason, one that our life of Darwin vividly illustrates. The only way that Darwin could rid his theory of God, was to ensure that our peculiar human and moral capacities were reduced to a level at which they could be entirely explained by natural selection; or to put it the other way around, making everything hang on natural selection was a way to eliminate the necessity of God in the theory. Thus, God could not be eliminated without eliminating humanity, i.e., reducing human beings to mere animals. This, it seems to me, is a kind of interesting, if roundabout, proof of a central biblical doctrine, that human beings are made in the image of God. Darwin felt he had to destroy the image so that he could eliminate God, and the result was—quite justly—that in his picture of man, he couldn't even recognize himself. By that I don't mean that he might not have fancied himself looking somewhat like his alleged simian ancestors, but that there was no place for his own personal high moral nature and sentiments in his theory (let alone his admirable intellectual powers). As for me, I shall always prefer a theory of evolution that can explain so great a man as Charles Darwin.

Acknowledgments

As always, I'd like to thank my family for allowing me the necessary hours holed up in the writing shed, and Harry Crocker for his judicious editing that turned an unwieldly and winding manuscript into a tightly-written book. I am also grateful to Sam Reeves, Bruce Schooley, and tothesource Foundation for kindly funding the book budget for the project, and to Scott Hahn and the St. Paul Center for generously supporting me during the writing.

Endnotes

Chapter 1

1 Anna Seward, *Memoirs of the Life of Dr. Darwin, Chiefly During His Residence in Lichfield, with Anecdotes of His Friends, and Criticisms on His Writings* (Philadelphia: Wm. Poyntell & Co., 1804), 69.

2 The Darwin family was fond of sprinkling the same names—Erasmus, Charles, and Robert—liberally over each generation.

3 Desmond King-Hele, *Erasmus Darwin: A Life of Unequalled Achievement* (London: Giles de la Mare Publishers Ltd., 1999), 103.

4 Charles Darwin, *The Autobiography of Charles Darwin, 1809–1882* (New York: Norton, 1969), 28–29.

5 Janet Browne, *Charles Darwin: Voyaging* (Princeton: Princeton University Press, 1995), 160–61.

6 Erasmus Darwin, *Zoönomia*, (New York: AMS Press, 1974), I.xxxix.iv.8, 505.

7   Charles Darwin, *The Autobiography of Charles Darwin, 1809–1882* (New York: Norton, 1969), 23.

8   Ibid., 28.

9   Quoted in Janet Browne, *Charles Darwin: Voyaging*, 23.

10   Charles Darwin, *The Autobiography of Charles Darwin, 1809–1882*, 44.

11   Ibid., 25.

12   Ibid., 21.

13   Desmond King-Hele, *Erasmus Darwin: A Life of Unequalled Achievement*, 370.

14   Charles Darwin, *The Autobiography of Charles Darwin, 1809–1882*, 49.

15   Janet Browne, *Charles Darwin: Voyaging*, 83.

16   Ibid., 86–87.

17   Ibid., 83.

18   The Reverend Seward, the father of Erasmus's biographer-to-be, Anna Seward, saw right through the cryptic motto, and immediately accused Erasmus of being a modern Epicurus (a significant charge, since it meant being hedonist and atheist, as well as an evolutionist). This sudden discovery brought Erasmus to hide away his views for another two decades before bringing them to light again in his publication of *Zoönomia* in 1794. Desmond King-Hele, *Erasmus Darwin: A Life of Unequalled Achievement*, 89.

19   For a reproduction of each see Desmond King-Hele, *Erasmus Darwin: A Life of Unequalled Achievement*, 88 and 358.

20   Charles Darwin, *The Autobiography of Charles Darwin, 1809–1882*, 48.

21 Quoted in Janet Browne, *Charles Darwin: Voyaging*, 68.

22 Charles Darwin, *The Autobiography of Charles Darwin, 1809–1882 *, 28.

23 Ibid., 56.

24 Ibid., 39.

25 Ibid., 32.

26 Ibid., 47–48.

27 Ibid., 56.

28 Quoted in Desmond King-Hele, *Erasmus Darwin: A Life of Unequalled Achievement*, 301–2.

29 Janet Browne, *Charles Darwin: Voyaging*, 9.

Chapter 2

1 Charles Darwin, *The Autobiography of Charles Darwin, 1809–1882* (New York: Norton, 1969), 57.

2 Ibid., 62–63.

3 Janet Browne, *Charles Darwin: Voyaging* (Princeton: Princeton University Press, 1995), 107.

4 Charles Darwin, *The Autobiography of Charles Darwin, 1809–1882*, 61–62.

5 Quoted in Janet Browne, *Charles Darwin: Voyaging*, 113.

6 Charles Darwin, *The Autobiography of Charles Darwin, 1809–1882*, 64–65.

Chapter 3

1 Janet Browne, *Charles Darwin: Voyaging* (Princeton: Princeton University Press, 1995), 152–53.

2 Charles Darwin, *The Autobiography of Charles Darwin, 1809–1882* (New York: Norton, 1969), 71–72.

3 Janet Browne, *Charles Darwin: Voyaging*, 156, 160.

4   Letter 148, Darwin, C. R. to Whitley, C. T., 15 November
     [1831]. Happily, much of Charles Darwin's voluminous
     correspondence can now be found online at The Darwin
     Correspondence Project.
5   Letter 150, Henslow, J. S. to Darwin, C. R., 20 Novem-
     ber 1831.
6   Quoted in Janet Browne, *Charles Darwin: Voyaging*, 178.
     These words were actually written near the *end* of his
     voyage, thus attesting to the fact that he was morbidly
     sick of being seasick.
7   Ibid., 237.
8   Ibid., 184.
9   Charles Darwin, *Voyage of the Beagle*, ed. Janet Browne
     and Michael Neve (London: Penguin, 1989), 41.
10   Aristotle, *Metaphysics* 982b12–14. Hippocrates Apostle
     translation.
11   Charles Darwin, *Voyage of the Beagle*, 42–43.
12   Ibid., ch. II, 62-63.
13   Charles Darwin, *The Autobiography of Charles Darwin,
     1809-1882*, 77.
14   Quoted in Janet Browne, *Charles Darwin: Voyaging*, 217.
15   Ibid., 218.
16   Charles Lyell, *Principles of Geology*, Vol. II, ch. 1,
     193–95.
17   Ibid., Vol. II, ch. 2, 200–3.
18   Ibid., 205–6.
19   Ibid., 207–9.
20   Ibid., ch. 3, 210–12.
21   Ibid., 217–19.
22   Charles Darwin, *Voyage of the Beagle*, ch. XI, 171.

23   Ibid., 172.

24   Janet Browne, *Charles Darwin: Voyaging*, 243.

25   Charles Darwin, *Voyage of the Beagle*, ch. XI, 172, 177–78.

26   Ibid., 177, footnote.

27   Charles Darwin, *Voyage of the Beagle*, ch. III, 72.

28   Ibid., 72–73.

29   Ibid., 73–74.

30   Ibid., 76.

31   Quoted in Adrian Desmond and James Moore, *Darwin: the Life of a Tormented Evolutionist* (New York: Norton, 1991), 141.

32   Desmond King-Hele, *Erasmus Darwin: a Life of Unequalled Achievement* (London: DLM, 1999), 2–3.

33   Quoted in Desmond King-Hele, *Erasmus Darwin*, 2.

34   Charles Darwin, *Voyage of the Beagle*, ch. XIX, 287.

35   Ibid., ch. XX, 293.

36   Ibid., 301–2.

37   Ibid., 305.

38   Ibid., 306.

39   Ibid., 309.

40   Ibid., 313–14.

41   Ibid., ch. XXI, 320–21.

42   Ibid., 323–24.

Chapter 4

1   Charles Darwin, *The Autobiography of Charles Darwin, 1809–1882* (New York: Norton, 1969), 78–79.

2   Letter 310, Darwin, C. R. to FitzRoy, Robert, 6 Oct [1836].

3   Quoted in Janet Browne, *Charles Darwin: Voyaging*, 344.

4   Letter 311, Darwin, C. R. to Henslow, J. S., 6 Oct [1836].

5   Quoted in Janet Browne, *Charles Darwin: Voyaging*, 354.

6   For the letter and Darwin's response, see Ibid., 411.

7   Paul Barrett, et al, eds., *Charles Darwin's Notebooks, 1836-1844: Geology, Transmutation of Species, Metaphysical Enquiries* (Ithaca, NY: Cornell University Press, 1987), Notebook B:18, 175.

8   Ibid., Notebook M:84e, 539–40.

9   Ibid., Notebook M:27, 31, 526–27.

10 Ibid., Notebook M:19, 524.

11 Ibid., Notebook B:36, 180.

12 Ibid., Notebook B:44, 181.

13 Quoted in Janet Browne, *Charles Darwin: Voyaging*, 402.

14 Ibid., 434.

15 Ibid., 452.

16 Ibid., 469.

17 Quotes from both Sedgwick and Darwin in Laura Snyder, *Reforming Philosophy: A Victorian Debate on Science and Society* (Chicago: University of Chicago Press, 2006), 189–90.

18 Adam Sedgwick to Charles Lyell, April 9, 1845. See John Willis Clark, *The Life and Letters of the Reverend Adam Sedgwick* (Cambridge: Cambridge University Press, 1890), Vol. II, 84.

19 Quoted in Janet Browne, *Charles Darwin: Voyaging*, 501.

20 Ibid., 541.

Chapter 5

1   Charles Darwin, introduction to *On the Origin of Species by Means of Natural Selection, or the Preservation of*

Favoured Races in the Struggle for Life (London: John Murray, 1859), 5 (Facsimile of the First Edition, Cambridge: Harvard University Press, 1964).

2   Letter 2269, Darwin, C. R. to Hooker, J. D., 6 May [1858].

3   Letter 2306, Darwin, C. R. to Hooker, J. D., 13 July [1858].

4   Charles Darwin, "Slave-making instinct," in *On the Origin of Species*, ch. VII, 219–24.

5   Quoted in Howard Gruber, *Darwin on Man* (New York: Penguin, 1974), 66.

6   Quoted in Janet Browne, *Charles Darwin: The Power of Place* (Princeton: Princeton University Press, 2002), 85.

7   Charles Darwin, *The Descent of Man*, Part I, ch. V, 162–63.

8   Ibid., ch. III, 103.

9   Ibid., ch. V, 168.

10 Ibid., 168–69.

11 Quoted in Adrian Desmond and James Moore, *Darwin's Sacred Cause: How a Hatred of Slavery Shaped Darwin's Views on Human Evolution* (Boston: Houghton Mifflin, 2009), 303.

12 Quoted in J. R. Lucas, "Wilberforce and Huxley: A Legendary Encounter," *The Historical Journal*, 22 (1979), 313–30.

13 Quoted in Adrian Desmond and James Moore, *Darwin's Sacred Cause*, 321.

14 It appears for the first time in the third edition. Here, I refer to the sixth edition, with an introduction by Julian Huxley (New York: Mentor, 1958), "Historical Sketch," 17.

15 Janet Browne, *Charles Darwin: The Power of Place*, 129.

16 Quoted in Adrian Desmond, *Huxley: From Devil's Disciple to Evolution's High Priest* (Reading, Massachusetts: Addison Wesley, 1997), 287.

17 Asa Gray, *Natural Selection not inconsistent with Natural Theology: A Free Examination of Darwin's Treatise on the Origin of Species and of its American Reviewers* (London: Trübner & Co., 1861), 52.

18 See Janet Browne, *Charles Darwin: The Power of Place*, 175–77.

19 Ibid., 219.

20 Alfred Russel Wallace, "Sir Charles Lyell on Geological Climates and the Origin of Species," *Quarterly Review* (April 1869), 391.

21 Ibid., 391–93.

22 Ibid., 393–94.

23 See Janet Browne, *Charles Darwin: The Power of Place*, 318.

24 Charles Darwin, *The Descent of Man*, Part I, ch. II, 35–42, 46, 52–57, 60–61, 62–64.

25 Ibid., 65–68

26 Ibid., 35.

27 Charles Darwin, *Voyage of the Beagle*, ch. XI, 177, footnote.

28 Charles Darwin, *The Descent of Man*, Part II, ch. XXI, 388–89.

29 Ibid., Part I, ch. VI, 200-1.

30 Ibid., ch. IV, 145–46.

31 Michael Polanyi, *Personal Knowledge: Towards a Post-Critical Philosophy* (Chicago: University of Chicago Press, 1962); and Thomas Kuhn, *The Structure of Scientific*

Revolutions, 2nd edition (Chicago: University of Chicago Press, 1970).

Chapter 6

1   St. George Mivart, *On the Genesis of Species* (London: Macmillan & Co., 1871), 4.
2   Ibid.
3   Ibid., 5.
4   These are first listed by Mivart, and then treated one by one in detail. See St. George Mivart, *On the Genesis of Species*, 21ff.
5   St. George Mivart, *On the Genesis of Species*, 37–38.
6   Charles Darwin, chapter VII, "Miscellaneous Objections to the Theory of Natural Selection," in *Origin of Species*, 6th ed. (Mentor edition), 204.
7   St. George Mivart, *On the Genesis of Species*, 24–26.
8   Charles Darwin, chapter VII, "Miscellaneous Objections to the Theory of Natural Selection," in *Origin of Species*, 6th ed., (Mentor edition), 206–7.
9   Ibid., 214–16.
10 St. George Mivart, *On the Genesis of Species*, 262.
11 Ibid., 288.
12 See Peter Bowler, *The Eclipse of Darwinism: Anti-Darwinian Evolutionary Theories in the Decades around 1900* (Baltimore: Johns Hopkins University Press, 1983).
13 See Chris Stephen and Allan Hall, "Stalin's half-man, half-ape super-warriors," *The Scotsman*, December 20, 2005.

Chapter 7

1   See the classic account by Frank Manuel, *The Eighteenth Century Confronts the Gods* (New York: Atheneum, 1967).

2   On the secular goal defining and driving the intellectual
     and cultural efforts, see Peter Gay, *The Enlightenment: An
     Interpretation. The Rise of Modern Paganism* (New York:
     Norton & Co., 1966), but also the more recent, and thor-
     oughly marvelous Jonathan Israel, *Radical
     Enlightenment: Philosophy and the Making of Modernity,
     1650-1750* (Oxford: Oxford University Press, 2001).
3   Frederick Burkhardt, et al, *The Correspondence of
     Charles Darwin*, XI.278.
4   Adrian Desmond and James Moore in their *Darwin's
     Sacred Cause: How a Hatred of Slavery Shaped Darwin's
     Views on Human Evolution* (New York: Penguin Books,
     2009).
5   Charles Darwin, chapter IV, "Natural Selection," in *Ori-
     gin of Species*, 6th edition (Mentor), 93–94.
6   See my *Ten Books That Screwed Up the World* (Washing-
     ton, DC: Regnery, 2008), ch. 11; Adolf Hitler, "Mein
     Kampf"; and Richard Weikart, *From Darwin to Hitler:
     Evolutionary Ethics, Eugenics, and Racism in Germany*
     (New York: Palgrave MacMillan, 2004).
7   Although not directed specifically at providing a broader
     and deeper foundation for evolutionary theory, I have
     sketched some of the most salient aspects that would have
     to be included in an evolutionary theory larger than Dar-
     winism in my *Meaningful World: How the Arts and Sci-
     ences Reveal the Genius of Nature* (IVP, 2006;
     co-authored with Jonathan Witt). The argument is too
     subtle, complex, and multi-layered to be produced in an
     outline form here, so I direct interested readers to the
     book.

Chapter 8

1   Charles Darwin, *The Descent of Man*, Part II, ch. XXI, 405.

2   Ibid., 403.

3   Ibid., Part I, ch. V, 177.

4   For more on these figures see my *Architects of the Culture of Death* (CO: Ignatius, 2004; co-authored with Donald DeMarco); and my *Ten Books That Screwed Up the World* (Washington, D.C.: Regnery Publishing, Inc., 2008).

5   Charles Darwin, *The Descent of Man*, Part I, ch. VII, 238.

6   Ibid., ch. VI, 201. Emphasis added.

7   Quoted in Desmond and Moore, *Darwin*, 591.

8   Ernst Haeckel, *The Wonders of Life: A Popular Study of Biological Philosophy*, trans. Joseph Mc Cabe (New York: Harper, 1905), 119–20; and Daniel Gasman, *The Scientific Origins of National Socialism: Social Darwinism in Ernst Haeckel and the German Monist League* (London: MacDonald, 1971), 32–36.

9   Quoted in the excellent Richard Weikart, *From Darwin to Hitler: Evolutionary Ethics, Eugenics, and Racism in Germany* (New York: Palgrave MacMillan, 2004), 2.

10 Ibid., 45.

11 The quote is from Nietzsche's *Ecce Homo*, "The Birth of Tragedy," Section 4. Nietzsche's German is "die schonungslose Vernichtung alles Entartenden und Parasitischen," which Weikart (p. 49) translates as quoted in the passage, and Walter Kaufmann, as "the relentless destruction of everything that was degenerating and parasitical," thereby taking the stress of persons and putting them on

things. (Friedrich Nietzsche, *On the Genealogy of Morals and Ecce Homo* [New York: Vintage, 1967].) I am inclined toward Weikart's translation, as Kaufmann is forever trying to distance Nietzsche from the real implications of his philosophy in the Third Reich.

12 See Richard Weikart, *From Darwin to Hitler*, 15.

13 See H. W. Crocker III, *Don't Tread on Me: A 400-Year History of America at War, from Indian Fighting to Terrorist Hunting* (New York: Crown Forum, 2006), p. 420; and Stephen Jay Gould, *Bully for Brontosaurus: Reflections on Natural History* (New York: W. W. Norton & Company, 1991), 424.

14 Quoted in Robert Lifton, *The Nazi Doctors* (New York: Basic Books, 2000), 31.

15 Adolf Hitler, *Mein Kampf* (Mumbai: Jaico Publishing House, 1988), Volume I, Chapter Two, 39.

16 Ibid., Chapter Three, 97.

17 Ibid., Volume II, Chapter One, 348–49.

18 Quoted in Richard Weikart, *From Darwin to Hitler*, 210.

19 Ibid., 227.

Index